Management for Professionals

For further volumes:
http://www.springer.com/series/10101

Clifford M. Gross

Too Good To Fail

Creating Marketplace Value
from the World's Brightest Minds

 Springer

Clifford M. Gross
Tekcapital Limited
Oxford Center for Innovation
Oxford
UK

ISSN 2192-8096   ISSN 2192-810X   (electronic)
ISBN 978-3-319-00280-4   ISBN 978-3-319-00281-1   (eBook)
DOI 10.1007/978-3-319-00281-1
Springer Cham Heidelberg New York Dordrecht London

Library of Congress Control Number: 2013939953

Printed on acid-free paper

Springer is part of Springer Science+Business Media (www.springer.com)

For Marielle, Harrison, Bradley, Linda and Ilyse

Preface

This small volume would not have been written had I not decided to return to school to freshen-up my thinking after several years in business and academia. To this end, I am indebted to my Professors and colleagues at Oxford's Saïd Business School; where many inspirational lectures and after-class discussions provided the basis for annealing my thoughts on technology transfer through the lens of network strategies, open innovation, and design thinking. On this note, special thanks are due to Professors Marc Ventresca, Sue Dopson, and Lucy Kimbell whose lectures on strategy and innovation, management and design thinking respectively, provided an illuminating reference frame from which to view many of my past experiences, both the failures and the successes.

I am particularly grateful for the unwavering support of my friends and colleagues at Tekcaptial provided while I wrote this text, especially Max Ingles, Arian Lewis, Anthony Mescall, Selwyn Lloyd, Dr. Jeffrey Bleil, Carl Nisser, Barry Grossman, John Borer, III, and Jeffrey Lewis. Without their help, many of the ideas discussed herein would have remained in the realm of quiet notions and never would have seen the light of day. I also owe a debt of gratitude to Paul Dorfman of the New York Stock Exchange, whose friendship, encouragement, and enthusiasm for open innovation provided steadfast inspiration.

Contents

Introduction

The empires of the future are the empires of the mind.

Winston Churchill[1]

Innovation results in the development of new technologies and business models to improve the world around us. From faster chips, the more efficient conversion of sunlight to electricity and better ways to diagnosis and treat diseases, innovation describes the inexorable improvement in the application of science and technology to enhance the quality of life and create lasting value. The myth of the $538 garage entrepreneur has since David Packard and a few other outliers, given way to long term, significant research infrastructure of university laboratories staffed with scientists, physicians, and engineers envisioning generation 2.0 of the current state of the art. While many companies have significant research and development facilities, most corporate R&D efforts are mainly D, or the improvement or application of the existing science and technology to enhance the design of products and services. The big research or the R in R&D is more often than not conducted at the major research universities and government-operated research laboratories. For these off-market or exogenous sources of innovation to be useful, they must be transferred to the companies that can commercialize them. Enter technology transfer and open innovation. Technology transfer is as the name implies, the art and science of relocation of pure research led innovation to commercial enterprises. It is a difficult, inefficient business due to the inherent mismatch in motivation and direction. Basic research pursues big ideas that can potentially address the major questions facing a particular field of inquiry, technology push if you will, whereas companies in one form or another are selling solutions to their customers. These solutions are in constant need of technological advancement to improve their efficacy and cost-effectiveness in solving customer problems. Technology is pulled from exogenous and local sources to inform these solutions. Technology transfer sits at the intersection of technology push and market pull. Historically, this mismatch was a fairly local phenomenon, with universities and research labs in rough proximity to corporate enterprises

---

[1] Sir Winston Churchill, Speech at Harvard University, September 6, 1943.

providing the lion's share of new discoveries to invigorate their corporate neighbors. The Internet changed the playing field by redefining local to global. Social media changed it again by interconnecting all sources of innovation to every point in the network. What was once exogenous and foreign has become ubiquitous, global and familiar. Strictly speaking, for companies there is only one major source of research, the global network of all research laboratories; most of which are located in universities. While this has simplified and scaled-up the supplier network, it has done nothing to address stress riser between technology push and market pull and the difficulty of getting these new discoveries to market.

Professor Henry Chesbrough coined the term Open Innovation in 2003, recognizing that for companies to be consistent innovators they would need to source new developments outside their four walls and apply their development competence to deliver them to their customers. The limited, controlled internal corporate R&D facility, exemplified by Bell Labs of old, was an endangered if not extinct species. Owning the majority of the bright minds in any field was no longer possible. Alan Lafley, P&G's former CEO and Chairman said it well "for every smart scientists we have in-house, there are approximately 200 equally smart scientists worldwide, toiling in the same exact discipline," hence, the need to be open to innovations that are developed outside your organization. Lafley ate his own cooking by establishing Connect and Develop at P&G to identify and acquire external developed innovations. It was a very good meal, doubling the market capitalization of P&G between 2000 and 2010; a seemingly impossible job a priori.

The next paradigm for accelerating the growth of companies is the convergence of global university innovation networks, not dissimilar from the Gaia Hypothesis, to provide consistent external innovation to business. This is of course easier said than done as universities and business are cut from different cloths and there are few tools and companies to facilitate transactions. As a result, world wide, approximately 80 % of university discoveries go un-licensed. These unutilized discoveries represent the fruits from some of world's brightest scientists, engineers, and physicians. Nature may abhor a vacuum, but she hates waste and routinely deselects it. For business to evolve, it must develop ways to consistently harness the scientific potential energy of the great universities of the world, whose discoveries are simply too good to fail. The improved productivity of companies and nations depends on it.

Chapter 1
Open Innovation and Intellectual Capital

Intellectual Capital: The World's Most Valuable Asset Class

Intellectual capital is the market value of an asset minus its book value. Strictly speaking it is the brain premium bestowed on physical and financial assets as a result of the ideas contained within those assets. Intellectual capital (IC) may be stratified to include human, structural, and relationship capital. Human capital is, as the name implies, the value that people add to an organization; structural IC is typified by patents, trademarks, copyrights, and organizational facilities and procedures such as Intel's ability to develop an improved computer processor. Relationship IC is described by the ways in which organizations deal with their customers in areas such as product development, sales, customer services, and responsiveness to feedback.[1] Social media companies such as Facebook, LinkedIn, and Twitter have high relationship IC and derive most of their market value and indeed revenue from their ability to monetize their networked customer relationships.

Intellectual properties (IP) consist of the "Intangible rights protecting the products of human intelligence and creation."[2] Intellectual property is important to business because it provides opportunities for growth by empowering an exclusive franchise, providing for defense against imitators, and presents the opportunity for both licensing and assertion against infringers. Most importantly, IP enables companies to offer products and services that are uniquely beneficial; enabling its customer to do things they otherwise could not do, or do things better, quicker, and less expensively. IP is the land upon which protected economic castles can be built or as famed investor, Warren Buffet has said, "In business, I look for economic castles protected by unbreachable moats."[3]

Intellectual property is first and foremost an asset class. An asset class is a type of property or a security underlying the property that behaves in a similar manner

---

[1] http://en.wikipedia.org/wiki/Intellectual_capital
[2] http://legal-dictionary.thefreedictionary.com/Intellectual+Property
[3] http://www.heraldtribune.com/article/20120315/article/120319731

C. M. Gross, *Too Good To Fail*, Management for Professionals,
DOI: 10.1007/978-3-319-00281-1_1,
© Springer International Publishing Switzerland 2013

1

in the marketplace and its use is governed by similar laws and regulations. Stocks, bonds, commodities, real estate, financial instruments like cd's, T-bills and of course patents, trademarks and, copyrights each describe an asset class.[4]

The chief goal of management is through the structure, strategy, and execution of the business; enhance the value of the assets a company holds and in doing so increase shareholder value. This is done by delivering products and services that create value for its customers. Innovation is a vital part of this value creation process and as Peter Drucker has noted, "Innovation is the specific instrument of entrepreneurship. The act that endows resources with a new capacity to create wealth."[5]

Most firms have a difficult time managing their intellectual properties due to their intangible nature and the often long delay between the development of IP and the resulting creation of market value. This would not matter much were it not for the fact that roughly 80 % of GDP growth may be attributed to development of innovations that allow us to do things better, faster, less expensively, or in some cases allow us to do things that were previously impossible, e.g., a two-way video conversation across the world with a mobile phone at almost zero cost.[6] Further, it has been shown that intellectual property accounts for roughly 40 % of the net asset value of all corporations in the U.S.[7] At the risk of stating the obvious, it is worth considering that brain power has become the most valuable asset in the world.

Innovation, like leadership, has always been and remains the wildcard in business success. This was the case in the 1800s with the emerging ice industry and is equally central in today's solar energy industry.

Cold Trends and How Markets for Natural Ice Emerged in the Nineteenth Century US[8]

The development of the market for natural ice in the nineteenth century was the culmination of a long-standing demand for preservation of food, processing meat, helping farmers and fisherman, and addressing beverage and medicinal applications. A well-defined articulated demand for portable ice was not on the menu, rather the general desire for enhanced food shelf life, better medical conditions for the infirmed, and cool, potable beverages. Demand aside, one could say the market for natural ice formally began with the patenting of the ice plow by Nathaniel Wyeth in 1825.[9] This seminal invention combined with dozens of incremental improvements, tools of the trade as it were, resulted in improved quality,

---

[4] http://www.investopedia.com/terms/a/assetclasses.asp#axzz2J7RURfR8

[5] http://www.brainyquote.com/quotes/authors/p/peter_drucker.html#Gl3PPgWY3IIyfss4.99

[6] Palfrey (2012). Referring to the work of Nobel Laureate Robert Solow.

[7] IBID from Idris (2003), p. 34.

[8] Ventresca (2012).

[9] Geroski (2009).

standardization, and reduced cost of natural ice. In Moore's Law-like fashion, the price per ton to of ice dropped exponentially between 1825 and 1860.

As with the development of many new markets there is an inexorable tug of war between technology push versus. technology pull.[10] Technology push describes the innovation developed by the producers of the technologies and culminated in the development of ice making machines, first patented by Jacob Perkins in London in 1834 and later with improvements by Dr. John Gorrie in Florida in 1851.[11] Technology pull is the incremental demand for improvements requested or desired by the market and usually limited to non-disruptive, Kaizen-like adaptations for quality enhancement and price reduction. These normally follow well-articulated marketplace desires under the existing product paradigm which according to Geroski, "supply push innovation processes are unlikely to produce a single new good or service." Innovation on the other hand, seeks to satisfy the imperfectly articulated demand with a specific, workable, and off times elegant solution.

Delivering these solutions requires the maturation of technology trajectories in line with consumer demand. In The Structure of Scientific Revolutions, Thomas Kuhn[12] defines a science as a discipline where the major questions are known and more or less agreed upon by the participants of the field. This has important implications for setting expectations of the scientists and engineers developing a new technology and directing the path that technology refinements, both incremental and disruptive, follow (Gross et al. 2000).

Companies, products, and technologies, like individuals, have life spans (Utterback 1994). The creative destruction that played out in the growth of the natural ice business and marketplace in the nineteenth century demonstrates this fact and serves as a poignant example of how suppliers and consumers of products advance together, in a non-Cartesian waltz, which improves the quality of life, while fostering commerce and wealth creation.

Bright Spots in Solar Energy

The solar energy industry has experienced rapid growth due to the escalating economic and environmental costs of fossil fuel generated electricity (Fig. 1.1). The major technology currently underpinning solar power electrical generation is photovoltaics; a method of generating electrical power by converting solar radiation into electricity. This is done using semiconductors that express the photovoltaic effect.

While conceptually quite simple, solar power generation uses arrays of photovoltaic materials to form panels. These panels generate DC current when exposed to

---

[10] IBID.

[11] IBID.

[12] Kuhn (1962).

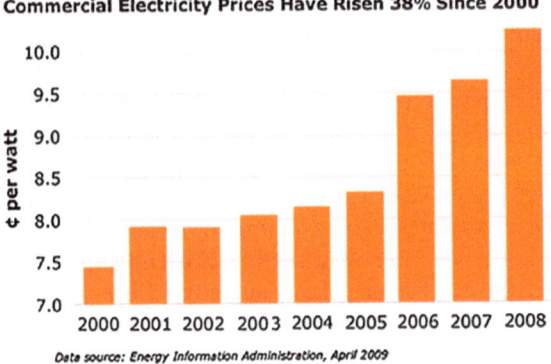

Fig. 1.1 Increase in the cost of electricity is driving the solar industry (Energy Information Administration 2009)

sunlight. For grid connectivity, the DC current must subsequently be converted to AC current using an inverter. The challenge of the photovoltaic industry is to produce this AC current at a price that is competitive with coal fired or nuclear powered power plants, an extraordinarily difficult challenge. Solar cells are currently made from monocrystalline silicon or polycrystalline silicon, and recently thin film cells from amorphous silicon, cadmium telluride, and copper indium gallium sulfide have provided alternative materials and production processes.[13] There are currently more than 15 different types of solar cells being produced by tier I manufacturers, worldwide (Energy Information Administration 2009).

The photovoltaic industry ecosystem is characterized by solar cell manufactures, systems integrators, and advanced technology licensing organizations. It is a well-developed industry with 39 public company participants with an average market capitalization of $3.67B.[14] Industry production of solar cells has risen dramatically on a global basis between 2001 and 2010 (Fig. 1.2).

From 2005 to 2009, world solar installed power has increased from 5 to 23 GW (Fig. 1.3).

These increased installed volumes have tracked reductions in cost per watt resulting from technological advances. The reciprocal of the typical S curve described increases in performance can be seen when installed cost is plotted versus time (Fig. 1.4).

In spite of this progress, the industry, while growing, is under great duress as it is highly reliant upon feed-in tariffs to help offset the higher costs relative to fossil fuel electricity production.

---

[13] Wikipedia (2012).

[14] Capital (2012).

Fig. 1.2 Solar cell production by region (Greentech Media 2012)

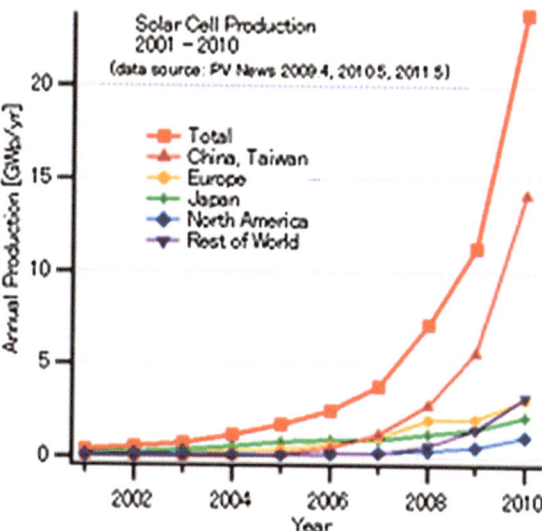

Fig. 1.3 Global photovoltaic power production (Martinot and Sawin 2009; European Photovoltaic Industry Association 2012)

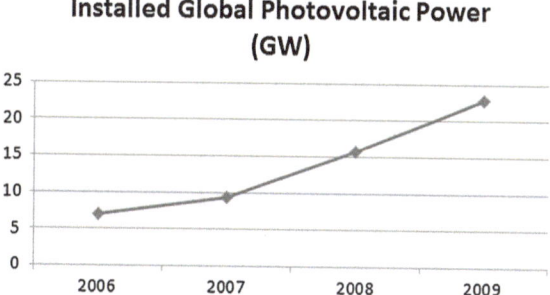

Photovoltaic Market Disruption and Value Creation

A unifying characteristic of the solar power industry is that all of the participants are focused on achieving electrical generation cost parity with nuclear, oil, gas, or coal fired power plants (Barbose et al. 2012 and Biello 2010). This battle is being waged on the semiconductor level with new technologies and process improvements, on the system integration level with durable, efficient, low cost installation techniques utilizing new materials and technologies for better solar capture, storage, and reduced maintenance costs, and finally on the organizational level with feed-in tariffs, R&D tax credits, and other government subsidies along with public equity market participation. A characteristic S curve (Fig. 1.5)[15] shows a break where a significant technological improvement has the potential to

---

[15] Geroski (2009) and Ventresca (2012).

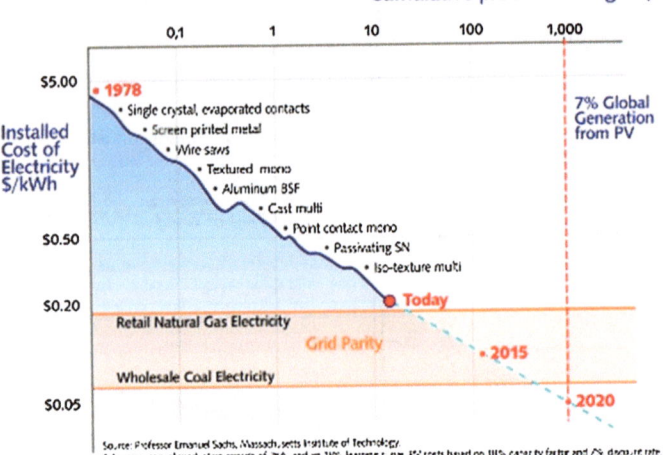

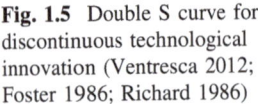

Fig. 1.4 The dropping cost of solar photovoltaics as a result of technological innovation (Geroski 2009; Rowse 2011)

Fig. 1.5 Double S curve for discontinuous technological innovation (Ventresca 2012; Foster 1986; Richard 1986)

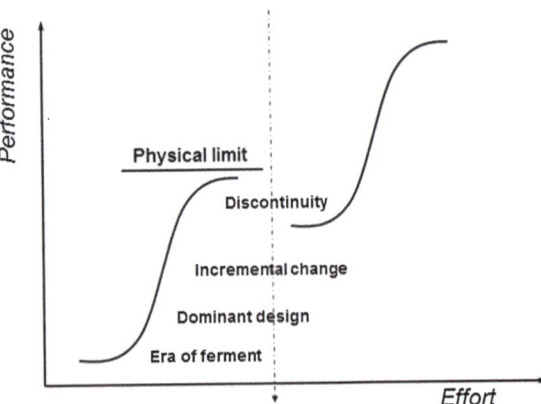

rationalize an industry. Casting molten silicon is just such an activity, creating a discontinuity in the technology trajectory and potentially fostering industry wide efficiency improvements. Currently, however, this is highly dependent upon one small company, 1366 Technologies, Inc. being able to execute, e.g., develop both an organization and successfully execute a go-to-market strategy. In general, durable marketplace disruption in the near-term is likely to arise from technology and process improvements that either dramatically increase the solar light to electricity conversion efficiency or reduced fabrication costs. Three potential, high value strategies for solar power marketplace value creation are as follows:

- Improvements to underlying PV technology via improved semiconductor design (e.g., in-licensing disruptive technology from a major research center).
- Improvements to PV manufacturing process to reduce cost of feedstock (develop or acquire process improvement technology for subsequent licensing).
- Improvements to company structure to enhance market penetration, via expanded staff competencies, improved brand awareness, and a superior balance sheet (e.g., JV, strategic alliance, or be acquired by a GE, Intel, or Siemens).

1366 Technologies, Inc. is an interesting, early stage, U.S. venture backed company (Kyle 2010) that has developed and patented a process improvement that is claimed to reduce the cost of PV module production (Wald 2010). The process involves casting the silicon wafers rather than sawing them off a large sheet. The Company claims that this may reduce the cost of cell production by 40 %. The research has been funded by venture investors and the U.S. Department of Energy. The initial work was conducted at MIT which makes this a technology transfer-based business. According to the Company, this may result in bringing the cost of solar produced electricity to $1.50/W. The Company had made additional technological improvements by perforating the cells with small holes which increase the surface area and result in enhanced efficiency through greater photon–electron interaction rates. In essence, 1366 Technologies is an R&D start-up with advanced process technology. Their core competency is leveraging university-based, scientific innovation focused on improved solar cell efficiency and cost reduction. They have no observable strength in go-to-market competency, manufacturing expertise, or known capability to leverage a global distribution channel. The board is academically focused and the balance sheet does not lend itself to meaningful OEM production or survivability for patent battles with entrenched suppliers, hence the technology push conundrum.

Creative Destruction

It is interesting to note that 1366 Technologies, Inc., notwithstanding their stated desire to becoming the world's leading supplier of low cost solar cells, is focused on manufacturing and supplying low cost cells to OEM's. This strategy in my judgment is likely to result in failure as more than 30 well financed competitors already exist and have invested significant sums in CAPX with government and/or capital market leverage. Schumpeter's creative destruction could likely come into play with the marginalizing of 1366 Technologies, Inc. as its first target, or as Schumpeter eloquently put it, "The promises of wealth and the threats of destitution that it holds out, it redeems with ruthless promptitude (Schumpeter 1942 and 1994)." The reason is straightforward, 1366 Technologies, Inc., while having many unique and positive attributes, probably does not have the go-to-market experience, manufacturing experience, or balance sheet to rationalize an industry with \$113B of market value—time will tell.[16]

A better approach, might be for 1366 Technologies Inc. to leverage the existing PV ecosystem with their novel and useful silicon foundry technology (intellectual capital), to be followed by additional R&D improvements, manifested in the form of high margin royalty streams. This could be achieved by ubiquitously licensing their technology to all photovoltaic manufacturers beginning for example, with a single major player with a two years exclusive, and then converting to a non-exclusive. This approach would create near-term value, and meaningful high margin revenue streams without embracing manufacturing and marketing risks. Further, through an effective licensing strategy, 1366 Technologies, Inc. could focus on its core competency of new process improvements and subsequently license these to OEM supplies. In the semiconductor space, Rambus has pursued this business model to good effect. Rambus develops and licenses its patented innovations, technologies, and architectures to digital electronics OEMs. As an innovation-driven company, Rambus' strategic focus is to develop technologies that enhance the end-user experience and create value for its shareholders through up-front payments and royalty streams. With this strategy, Rambus has achieved a current market cap of \$876M. Fittingly, the old BASF advertising saw "We don't make the chip, we make it better," is highly apropos. The point being that innovation alone is usually not sufficient to create a market leader; the technology must be coupled with an executable business model.

This is consistent with a strategy to demarcate a market with the development of alliances with industry stalwarts with revenue sharing agreements (Fig. 1.6). The overarching goal is to blend structural and relationship intellectual capital into a formidable, competitive advantage.

Elements	Claiming a Market	Demarcating the Market	Controlling the Market
Domain of actions	Cognitions	Relationships	Resources
Objective	Become cognitive referent in a distinct market.	Determine market perimeter, and define industry structure and roles for powerful players.	Cover market space, and eliminate or delay rivalry.
Dominant logic	Sensegiving	Co-optation	Ownership
Organizational capability	Shaping and promoting identity	Developing alliances	Making acquisitions
Mechanisms	Templates	Equity investments	Eliminate competing models
	Leadership signals	Revenue-sharing agreements	Increased market coverage
	Stories	Antileader positioning	Block entry of established firms

Fig. 1.6 Boundary strategies for market demarcation (Santos and Eisenhardt 2009)

---

[16] Capital IQ (2012).

The Take Away

In a fearlessly competitive marketplace, companies must seek disruptive innovations to accelerate and consolidate their position along the technology S curve. The plot thickens considerably when world beating technology requires masterful go-to-market execution, which it always does. The ice industry of the last century was revolutionized by Dr. Gorrie's invention of air conditioning, which notably evolved from his public health service work to treat patients with yellow fever and for which he received a patent in 1851.[17] The photovoltaic industry is in flux with technologies similar to 1366 Technologies' ability to cast silicon becoming potentially a disruptive innovation that empowers customers and an industry to produce electricity cost-competitively with burning coal. Interestingly, both of these corporate successes, although separated by more than 150 years, have resulted from leveraging exogenous innovation from federal and university laboratories.

What has changed in a century and a half is the pace of innovation. What took twenty years to evolve now occurs in two or three and it is continuing to accelerate. This speeding-up of technological innovation has rendered traditional, in-house R&D wholly inadequate in most instances to maintain the pace of innovation, let alone push the frontier of science and technology deterministically.

References

Barbose G, Darghouth N, Wiser R, Seel J. Tracking the Sun IV, an historical summary of the installed cost of photovoltaics in the United States from 1998 to 2010, September 2011. Lawrence Berkley National Laboratory, http://eetd.lbl.gov/ea/emp/reports/lbnl-5047e.pdf. Accessed 9 Feb 2012.

Biello D. Sunshine is free, so can photovoltaics be cheap? Sci Am. 2010. Accessed 19 Oct 2010.

Capital IQ. PV Industry Analysis, public companies. 2012. Accessed 8 Feb 2012.

Colville F. What is the real PV technology roadmap? PV tech, http://www.pv-tech.org/guest_blog/whatistherealPV technologyroadmap. 2012. Accessed 10 Feb 2012.

Energy Information Administration: http://visitsunworks.com/energy-price-protection.php (2009). Accessed 10 Feb 2012.

European Photovoltaic Industry Association: Market report 2011, http://www.epia.org/ (2012). Accessed 10 Feb 2012.

Foster FN. Innovation: the attacker's advantage. Summit Books; 1986.

Geroski PA. The evolution of new markets. Oxford University Press; 2009.

Gross C, Reischl U, Abercrombie P. The new idea factory, Columbus, Ohio: Battelle Press; 2000.

Idris K. Intellectual property: a power tools for economic growth. Geneva: WIPO; 2003. p. p34.

Kuhn TS. The structure of scientific revolutions. Chicago: University of Chicago Press; 1962. ISBN 0-226-45808-3.

Kyle A. 1366 raises $20M in Series B. Boston Bus J. 2010. Accessed 19 Oct 2010.

---

[17] http://www.phys.ufl.edu/∼ihas/gorrie/fridge.htm.

Martinot E, Sawin J. Renewables: global status report 2009 update, renewable energy world. http://www.renewableenergyworld.com/rea/news/article/2009/09/renewables-global-status-report-2009-update (September 2009). Accessed 10 Feb 2012.

Palfrey J. Intellectual property strategy. Cambridge: MIT Press; 2012.

PV NEWS (Greentech Media): http://www.greentechmedia.com/research/pv-news (2012). Accessed 10 Feb 2012.

Rowse B. Plummeting solar PV prices: a sustainability game changer, Carbonetix. 2011. http://www.carbonetix.com.au/news/plummeting-solar-pv-prices-a-sustainability-game-changer/. Accessed 10 Feb 2012.

Santos F, Eisenhardt K. Constructing markets and shaping boundaries: entrepreneurial power in nascent fields. Acad Manage J. 2009; vol 2(54).

Schumpeter JA. Capitalism, socialism and democracy. London: Routledge; 1942 and 1994.

Utterback JM. Invasion of a stable business by radical innovation: natural ice industry. Mastering the dynamics of innovation chapter 7. Boston, MA: HBS Press; 1994.

Ventresca M. Class handouts, strategy and innovation, EMBA 8. Oxford University; 2012.

Wald M. A cheaper route to solar cells, New York times. 2010. http://green.blogs.nytimes.com/2010/10/19/a-cheaper-route-to-solar-cells/. Accessed 10 Feb 2012.

Wikipedia: http://en.wikipedia.org/wiki/Photovoltaics (2012). Assessed 10 Feb 2012.

Chapter 2
The Technology Transfer Ecosystem

Following the great recession of 2008, there was a significant increase in unemployment in the United States. As a result, the U.S. Federal Reserve pursued with renewed vigor, its dual mandate of maximum employment and stable prices[1] as required by the Federal Reserve Act of 1913 and its subsequent amendment in 1977. The vibrancy of the U.S. economy in the 1990s with unemployment around 4 % and annual budget surpluses seems almost phantasmagoric, a distant memory. Most of the new jobs created in the 1990s and in fact for the last 30 years in the U.S. were created by small businesses.[2] Recently, the U.S. Congress approved the Jobs Act—or Jumpstart Our Business Startups Act, which was approved by the Senate 73–26, with some additional investor protections on March 22, 2012, to help repair the broken ecosystem that finances small public companies in the U.S. However well intended the Jobs Act, without a fully functioning ecosystem for new technology commercialization, or specifically tax relief for business willing to embrace additional go to market risk incumbent with new technologies, coupled with a streamlined patent system, the market failure is unlikely to be corrected.

Problems in the small cap equity market and the U.S. Patent and Trademark Office have contributed to the current employment malaise by synergistically reducing the commercialization of new technologies. To address this bifurcated problem and create a useful syzygy, a new tech transfer 2.0 model is needed to accelerate small company and job growth in the U.S. and other countries with similar underlying dynamics. The underlying mechanism is to facilitate the harnessing by business of science and technology innovation; the knock-on effect of which would be increased productivity.

---

[1] Mishkin (2007).
[2] Schramm (2009) and Weitekamp (2010).

C. M. Gross, *Too Good To Fail*, Management for Professionals,
DOI: 10.1007/978-3-319-00281-1_2,
© Springer International Publishing Switzerland 2013

Assessment of Technology, Market, and Organizational Capabilities

The technology, market, and organizational factors affecting the success of small companies in the U.S. are inextricably interwoven. A deficiency in one factor has the knock-on effect of impeding corporate growth and nationwide employment. The overarching theme is that a systems building framework is necessary to visualize and repair what is an obvious market failure.

The Patent Office

Innovative technology is vital for the growth of companies, industries, and the economy. Writ large, the inexorable advance of technology is reflected in the patent activity of firms, industries, and countries. Patents, while not innovations themselves, serve as a proxy for invention and the seed corn for innovation and job creation. According to the World Intellectual Patent Organization, on a global basis, patent applications are highly correlated to GDP[3] (Fig. 2.1). Between 1995 and 2011, the number of patent applications filed globally more than doubled from approximately 1 to 2.1 M per year. During this period, granted patents increased from 500 K per year to approximately 1 M per year, roughly tracking about 50 % of patent applications filed.[4] In 2011, China became the largest patent office in the world, having received more patent applications, utility models, industrial designs, and trademarks than the U.S. China surpassed Japan in this regard in 2010. In short, the world's factory has become the world leader in intellectual property production.[5] This will be discussed more fully in Chap. 7.

Although U.S. patent filings have continued at a good clip, their rate of issuance has slowed. There was a particularly large run-up in patent activity during the tech boom of the 1990s. This followed significant capital investment into new technology companies that leveraged the emergence of the Internet and information technology.

However, innovation is more than just the number of patents issued, as it also requires the commercialization of the invention i.e., capital plus a go-to-market strategy and execution, to result in productivity improvements.[6]

After 1999, the number of patents filed began to increase much faster than the number of issued patents (Fig. 2.2). This is problematic, especially for small businesses, which often rely of new patent issuances to raise equity financing.

---

[3] http://www.wipo.int/ipstats/en/wipi/figures.html#overview

[4] http://www.wipo.int/export/sites/www/ipstats/en/wipi/graphs/a_1_1_1.gif

[5] http://www.wipo.int/export/sites/www/ipstats/en/wipi/pdf/941_2012_highlights.pdf

[6] Seidel (2012).

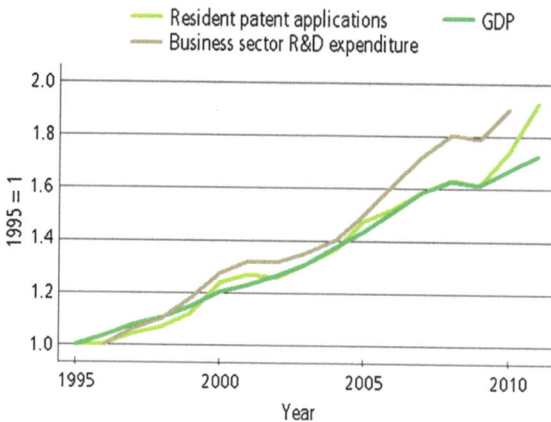

Fig. 2.1 The relationship between GDP, patent applications, and R&D expenditure (worldwide). *Source*: WIPO, World Intellectual Property Indicators 2012

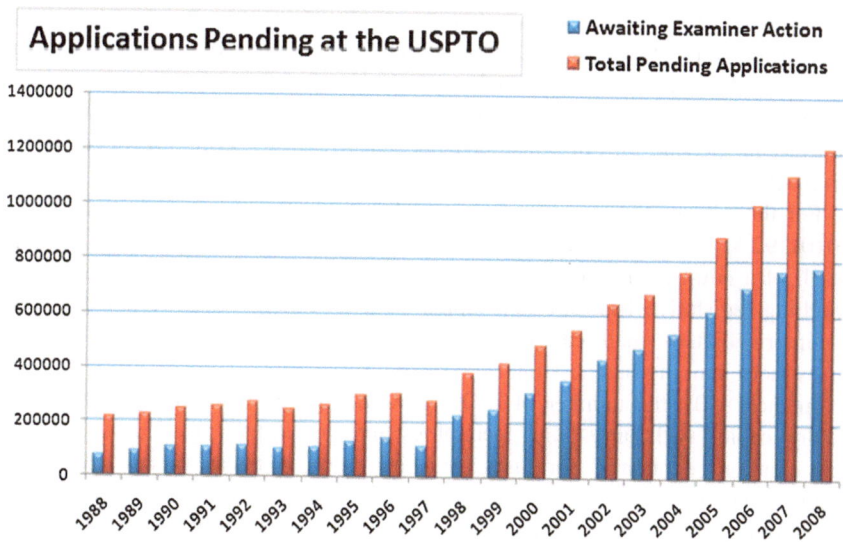

Fig. 2.2 Patents awaiting USPTO examiner action (Quinn 2011)

Nothaft (2010) believes that the patent office delay is due to a reduction in funding for the USPTO.

Nothaft (2010) estimated that approximately one million patent applications are currently awaiting examination by the USPTO. They have also estimated that three to ten new jobs are created for every newly issued patent, and that as a result approximately 2.5 M new jobs would be created if the USPTO review backlog were resolved. The current backlog is approximately three years from date of filing. President Obama's (2010) remarks on this are pretty telling, "Believe it or not, in our patent office—now, this is embarrassing—this is an institution responsible for protecting and promoting innovation—our patent office receives more than 80 % of patent applications electronically, then manually prints them out, scans them, and enters them into an outdated case management system. This is one of the reasons why the average processing time for a patent is roughly three years. Imminently solvable; hasn't been solved yet."[7] The usefulness of patents grows exponentially in the context of a well-financed corporate actor; hence, the need for linkage to corporate finance and tech transfer organizations as well as the public equity markets.

The Small Public Company

For small public companies (those with a market cap < $1B) to grow and prosper, technological innovation is paramount. Technologic innovation in small companies relies on many factors principally the availability of innovations, a capable team that can formulate and execute a go-to-market strategy and growth capital. Worldwide, universities are the largest source of innovative discoveries in the form of technologies available for license.

In the U.S., small firms add roughly three million jobs in their first year, a stark contrast to the one million jobs shed per year by larger firms (Weitekamp and Pruitt 2010).

For those firms that are able to access the public equity markets (Levensohn 2010), 92 % of job growth occur post-IPO. Currently, this is not happening as a result of a number of factors including the unintended consequences of an over-reaching legislative framework, coupled with technological advances that have dis-intermediated small cap brokerage firms rendering the small end of the investment banking business uneconomical.

In the 1990s, the number of IPO's grew rapidly as investors was hungry to deploy capital with early stage, technology enriched firms (Fig. 2.3). A period of job creation, economic surpluses, and increased business process efficiency resulted. These benefits came with the increased downside risk of tortious activities from firms like ENRON and WorldCom, whose bankruptcies inspired numerous financial acts designed to stem future abuses. Regulation FD was enacted to separate investment bank research from trading activities and additional rules e.g., Sarbanes–Oxley, were promulgated to ensure better financial reporting and controls. Also, improved Internet technology allowed investors to buy and sell stocks directly, without the direct assistance of a broker. This reduced both the importance of stock brokering and the income such activities generated for the brokerage houses.

---

[7] Obama (2010).

The IPO market is broken
In the last decade the number of IPOs has fallen dramatically, specifically deals less than $50 million in proceeds.

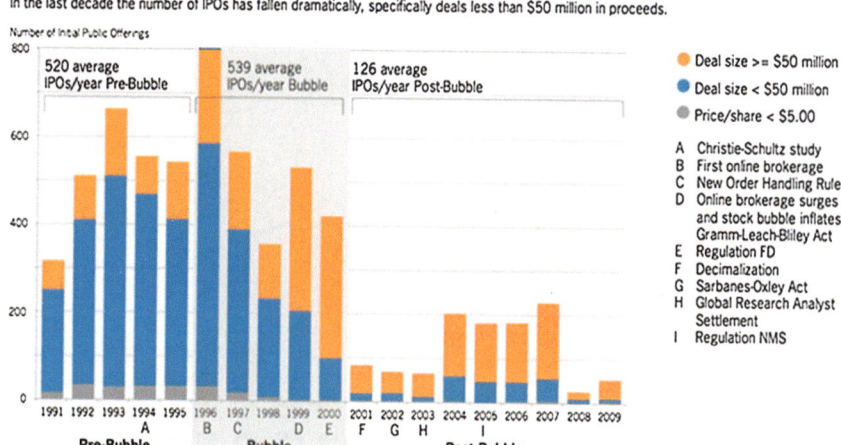

Fig. 2.3 IPO's and small cap market failure (Weild and Kim (2010)—based on data from Grant Thornton, LLP)

Stock quotes were decimalized from 1/16th of a dollar to 1/100 of a dollar (6.25 cents to one cent) to reduce bid-ask spreads. These narrower spreads resulted in reduced profits for market makers, which led to reduced liquidity as there was less capital available to invest.[8]

These factors certainly reduced the market for new IPO's post 2000 (Ritholtz 2009). Small cap IPO's suffered the most (IPO's raising less than $50 M). Many small cap banks closed-up shop, as they could no longer earn enough money by underwriting the small issues with their increased overhead and liabilities. Further, on-line stock trading and REG FD reduced their income generation potential. The IPO market was decimated, especially for the smaller offerings (Fig. 2.4). Currently, only a handful of small cap investment banks are left in the U.S. (Borer 2012).

As most new job growth is in small companies, the reduction in IPO's hit job creation and GDP growth hard. In 2011, there were less than 150 IPO's in the U.S., with an average size of $265 M (The Business Review, 2012), much less than the 520 per year needed to support 3 % GDP growth (Fig. 2.3).

Most public companies in the U.S. are small companies with less than 500 employees. These companies currently have a very difficult time raising capital in follow-on offerings. Without being able to access equity based, growth capital, it is nearly impossible for most of these firms to push their products and services up the technology S-curve (Seidel 2012) or even keep-up with better financed, larger competitors.

---

[8] Serchuk (2009).

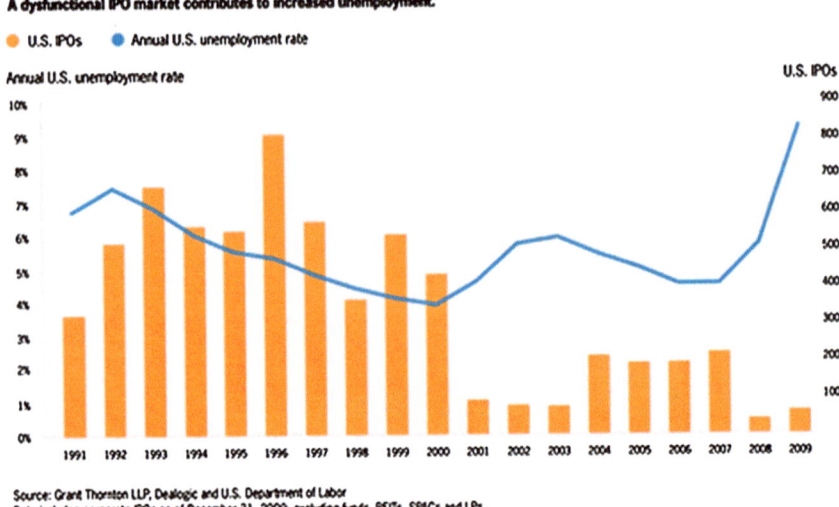

Fig. 2.4 IPO's and unemployment (Weild and Kim 2010)

Weild and Kim[9] have analyzed the effects of the post dot com regulatory and technological climate on the small cap market for both NASDAQ and NYSE listed firms (Fig. 2.5). This has contributed to market failure due to an in-virtuous network cycle.[10]

Of course, the post.com bubble and the subsequent subprime housing crisis also mitigated investor sentiment away from stocks as an asset class. Over the last 10 years, trading volumes on the NYSE are roughly half what they were a decade ago, and roughly half or more of the current trading volume is from computerized, high-frequency trading as compared with a more traditional equity investor.

The IPO market is the canary in the mine and sets the tone for the capital markets overall. The collateral damage of the IPO market downturn has debilitated the follow-on equity market, especially for companies with < $500 M in market capitalization. This has had the knock-on effect of forcing these smaller public entities to dramatically reduce overhead. When survival takes center stage, usually the first to go are R&D expenditures. As a result, the path for product and service improvement in these small public companies is hampered resulting in the downward spiral of dated products. Over time, competition tends to reduce a company's competitive advantage and its returns on invested capital. To remain competitive, a company must continually increase its proprietary technology to create long-term value. To help neutralize this weakness, it is vital that companies embrace open innovation (Chesbrough 2003) and seek to create a technology pipeline external to their business. This activity is squarely in the purview of

---

[9] Weild and Kim (2010)

[10] Seidel (2012).

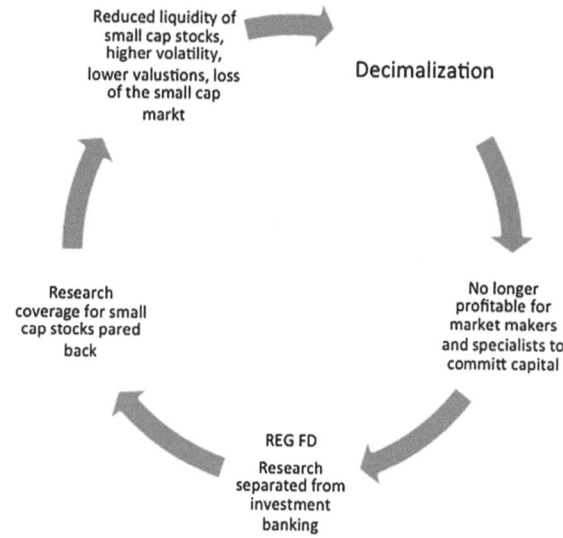

Fig. 2.5 The effect of decimalization and REG FD on the small cap market (After data from Weild and Kim 2010)

technology transfer firms or technology transfer operating groups within larger firms and can be a dominant strategy for creating a competitive advantage from discontinuous innovation (Ventresca 2012).

Technology Transfer: The Supplier Network

Technology transfer is the business of transferring technologies developed at one organization to other organizations to facilitate their commercialization. The typical process of technology transfer from the lens of the university is outlined in Fig. 2.6. In the U.S., there are more than 2,000 universities and 700 federally funded research laboratories (e.g., The Jet Propulsion Laboratory, Los Alamos National Laboratory, etc.) where government funded research is conducted. On December 12, 1980, the passage of the Bayh-Dole Act (P.L. 96-517, Patent and Trademark Act Amendments of 1980) granted universities and federal laboratories with property rights to these innovations and required that they seek to license or sell their discoveries that have been created with taxpayer funded research. As a result of Bayh-Dole, universities and federal laboratories were required to establish technology transfer offices to merchandise their innovations.

Prior to 1980, inventions developed by universities with taxpayer funded research were usually deemed to be in the public domain. As a result, most property rights could not be transferred and licensing was impractical. After the passage of Bayh-Dole in 1980, technology transfer from federally funded universities and research laboratories to commercial enterprises was established as an emerging industry.

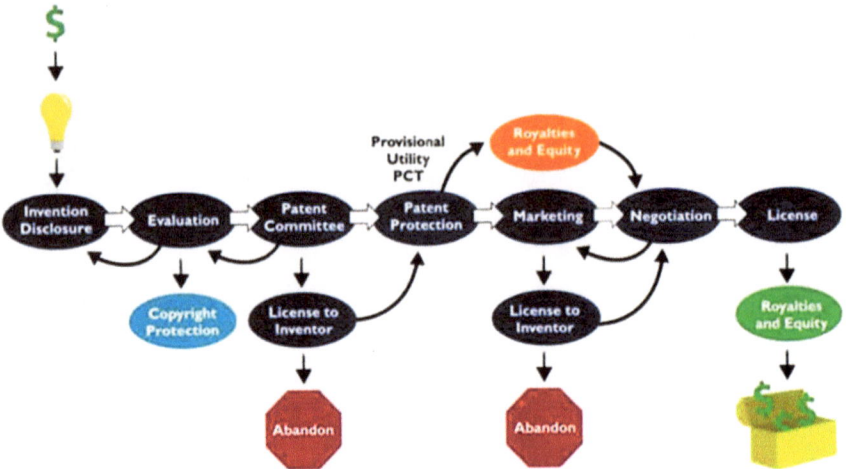

Fig. 2.6 The university technology transfer process (University of Toledo 2012)

In the U.S., the Association of University Technology Managers is the national organization that seeks to promote best practices in technology transfer. Similar organizations have been set up around the world, such as Praxis-Unico in the UK, to help develop the technology transfer ecosystem. Each year AUTM conducts a survey of the major research organizations to ascertain their success at transferring their technologies. The results of the 2010 annual licensing survey indicate that: "There were 657 new commercial products created, 4,284 licenses executed and 651 startup companies formed, 498 of which had their primary place of business in the licensing institution's home state. Expenditures for sponsored (government funded) research continued to increase from 2009 levels; total research and development spending increased 9.6 %, federal expenditures increased a robust 17.3 %, and industrial spending increased 5.6 %. A total of $59.1 billion total sponsored research expenditures with $39.1B coming from the federal government and $4.3B coming from industry. The number of invention disclosures received in 2010 was 20,642. Total license income was: $2.4 billion."[11]

Although there have been a number of stellar success in technology transfer such as gene splicing (licensed to Genentech which empowered the biotech industry) and Google (also from Stanford University), the Association of University Technology Managers estimates that in 2010, 80 % of university disclosed innovations went unlicensed.

As it currently stands, only 20 % of university discoveries find their way to the marketplace; the remainder languish. While unfortunate and a terrific waste of intellectual capital, this represents an opportunity to leverage the $47B spent annually on unlicensed university research products to exogenously assist

---

[11] AUTM (2011).

companies in the acquisition of externally developed innovation to power their business, a process known as open innovation.

To achieve this requires expertise in technology transfer coupled with the capital necessary to facilitate transactions and help catalyze their commercialization post-transfer.

The Technology Transfer Firm

According to Gross,[12] university technology transfer is a potential engine room for small cap growth by enabling companies to acquire new discoveries using an open innovation platform. Some of the more successful U.S. university technology transfers have been:

- Google, Stanford $193bn Mkt. Cap
- Genentech (Recombinant DNA), Stanford $46.8bn Mkt. Cap
- Gatorade, University of Florida acquired by PepsiCo ($13bn) as part of Quaker Oats Company acquisition
- Netscape (Mosaic-first internet search engine), University of Illinois $4.2bn acquired by AOL
- Akamai, MIT, $5.3bn Mkt. Cap
- Lycos, Carnegie Mellon $5.4bn acquired by Terra Networks, and
- Taxol, University of Florida, Cancer therapy, Bristol-Myers Squibb sales in excess of $1bn

From a Schumpeterian perspective, each of these examples can be viewed as having produced "creative destruction"[13] in their respective markets. These remarkable examples of university patented technology, brought to the marketplace, have resulted from the forward thinking Bayh-Dole Act of 1980. Since the enactment of Bayh-Dole in 1980, university patents have grown nearly exponentially (Fig. 2.7).

Fig. 2.7 The growth of university patents (The Tech—Online Edition 2010)

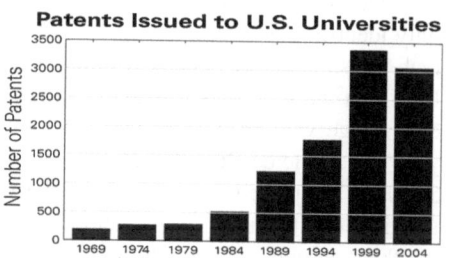

Patents Issued to U.S. Universities

---

[12] Gross (2003).

[13] Utterback 1994 as quoted in Ventresca (2012).

The U.S. has the largest university and federal laboratory infrastructure in the world for creating new IP (Gross 2003). With more than 2,000 colleges and universities and hundreds of national laboratories such as Los Alamos, NASA and the National Institutes of Health, etc.

The goal of technology transfer is to convey these technologies to companies that can commercialize them. For small companies, universities are often the best source for exogenously developed, potentially disruptive, and new discoveries; as they usually have modest abilities to conduct R&D themselves and R&D conducted by larger companies tends to stay within those firms.

Technology Transfer Competitive Landscape

The tech transfer market is fragmented and immature, largely because it relies on a very inefficient model of technology push.[14] It is supply driven, deep and long on the development of both new technologies and occasionally entirely new technology trajectories (Geroski 2009), such as gene splicing, which empowered the development of the entire bio-tech industry and Mosaic for X which led to the development of the browser and the commercialization of Internet. These technologies and the post hoc trajectories that incorporate them are in constant search for a customer or a company to commercialize them. Not much of a way to run a railroad.

Technology transfer organizations can be roughly classified into three groups:

• Tech transfer offices at major research centers,
• Venture capital and small cap investment banks
• Tech transfer consultancies and a mélange of consultancies that also invest in the spin out and commercialization of technologies.

The majority of all tech transfer activities occur within tech transfer offices. In the U.S., Canada, and the UK, venture cap firms occasional shop the major research centers for marketable discoveries and a small but insignificant number of transactions are consummated by tech transfer stand-alone consultancies. We have only identified one company in the U.S. that currently pursues the combination of tech transfer services and investment banking as a core business (MDB Capital). UTEK in the U.S. previously had an active tech transfer business combined with early stage venture investment. It achieved significant growth through 2006, the year it completed 26 transactions, and reached a market cap of approximately $200 M. It suffered, however, from the vicissitudes of the market by holding the acquired equity stakes for investment. As a result of the financial crisis in 2008, in 2009 it largely exited the tech transfer business and pursued innovation

---

[14] Geroski (2009).

management consulting engagements (it acquired several firms to do this including Gary Hamel's Strategos) as its core business.

In the UK, IP Group and Imperial Innovations both execute a capital intensive, combined model of tech transfer and early stage investment. Both companies could roughly be categorized as early stage venture funds with a university IP sourcing bias. Sagentia, PLC (formerly Scientific Generics) in the UK previously used this model but exited the business in 2006 to focus on their less risky and more predictable engineering consulting services, after losing 90 % of its market capitalization. BTG, the UK technology transfer privatization deal also pursued a similar strategy, proceeded with a flotation, lost most of its market capitalization and ultimately repositioned itself successfully as a life science investment and commercialization organization.

The vast majority of tech transfer offices are essentially not-for-profit arms of universities that seek to create marketplace value from in-house inventions. They do not, however, provide the necessary financing for the small companies to productize and market their innovations. They remain idea factories seeking marketplace champions.

A five forces analysis of tech transfer (Fig. 2.8) segments the fragmented tech transfer ecosystem. The buyers are mostly small cap companies (both private and public), the suppliers are the universities (they provide technology licenses), the investment banks and venture funds (provide access to capital) and the substitutes are the buyers themselves accessing capital from the equity and/or debt markets and in turn developing technologies.

An analysis of the competitive forces (Fig. 2.9) indicates some of the challenges tech transfer firms face.

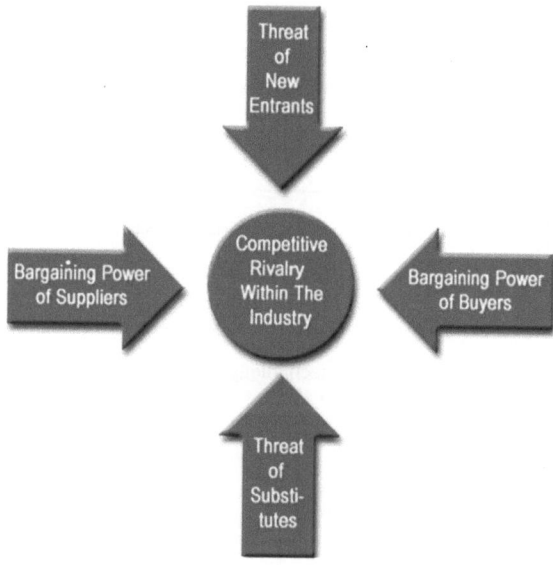

Fig. 2.8 The five forces describing the technology transfer industry dynamics (After Porter 1979 and from Investopedia.com 2012)

Potential Competitors/ Barrier of Entry	1.	Small companies are not loyal to existing technology transfer firms of any type as they have little or no experience working with them
	2.	Few acceptable (i.e., technology plus capital) substitutes are available due to the difficulty of small companies in raising capital
	3.	New entrants need both tech transfer and venture experience
	4.	Access to technologies available for license is non-trivial and requires amassing and screening global databases with sagacious selection

| Threat of Substitutes | 1. | No close substitutes are available that combine technology sourcing with capital |

Intensity of rivalry among established firms	1.	Tech transfer firms are their own worst enemies by pursuing a technology push strategy. Competitors don't materially worsen the situation
	2.	The industry is highly fragmented
	3.	The technology transfer industry is under pressure due to the financial crisis and the difficulty in assessing capital to commercialize new discoveries
	4.	Tech transfer offices at universities have fixed capital and organizational structures that prevents them from aggressively marketing their products by not having sufficient incentive compensation programs
	5.	IP Group and Imperial Innovations are well financed and have the knowledge to rationalize the technology push + capital space in the UK (while maintaining risky equity portfolios)
	6.	30 years after Bayh-Dole, tech transfer is still a word-of-mouth industry with little brand identity and market power with the exception of Stanford, MIT, Oxford and WARF (Wisconsin Alumni Research Foundation)

Bargaining power of buyers	1.	Small cap buyers are desperate for innovation + capital
	2.	Few companies are available to provide the service
	3.	One time purchase for most buyers
	4.	Difficult to switch suppliers because there are so few to switch to
	5.	Buyers cannot and do not want to become suppliers as it is non-core to their businesses

Bargaining power of suppliers	1.	Universities are durable, and can choose with whom they affiliate
	2.	Corporate licensees are dispersed and not vital to a given university
	3.	Switching cost are non-existent

Fig. 2.9 Analysis of the technology transfer marketplace

Fig. 2.10 The need to migrate network models in tech transfer from the college to the purposeful strategic network model

Tech transfer brokerage firm

Networks and the Nascent Technology Transfer Market

The invisible college network strategy[15] is currently used by the majority of the immature players in the tech transfer marketplace. From the perspective of this network model, universities are the suppliers and firms (mostly small public and private entities) are the buyers. However, the current push model does not work well, as evidenced by the relatively few transfers being consummated, in all jurisdictions. A new, deterministic, strategic brokerage network model (Fig. 2.10) is needed to facilitate transactions and help flip the market from technology push centric to market pull centric.

The majority of the existing tech transfer brokerage firms have immutably strong ties with their home institutions that normally preclude them from sourcing discoveries from other institutions that might be more helpful to their clients. The holes in the university networks are the weak ties with their client companies (unless they are spin outs). Stand-alone consultancies do not have this problem, they are technology agnostic and usually have weak ties with their suppliers and financers by fiat, the former of which in this market is a strength. However, they have the weakness of not having ownership of the technologies that a university tech transfer office has and so must manage a discovery network of universities alongside their strong ties, customer base. Gross[16] has described a market driven technology transfer model that begins with the company that needs the technology and has strong ties with tech transfer firm but little else. The firm is market driven as are its technology needs.

The small company can therefore readily leverage the tech transfer firm's weak ties to its university network to source and finance the technologies it needs to grow.[17] For tech transfer firms, weak ties are preferred to strong ties as they increase the selection of available technologies and in doing so reduce adverse selection. This simplistic process basically reverses the normal university technology push model, in favor of market pull by the small tech firms or small tech groups within larger firms. A key weakness however remains. Innovation requires

---

[15] Ventresca (2012).

[16] Gross (2003).

[17] Granovetter (1973).

more than technology, it also needs capital to commercialize it; hence, the need to develop connectivity between tech transfer firms and the capital markets.

To develop the market position, a successful technology transfer company would need to take the authoritative high ground with an enhanced approach for bringing participants together that reduces friction and facilitates closing transactions. This needs to be followed by conceptualizing and executing a business model that creates real value for market suppliers and buyers by compressing time to market.

Finally, an incumbent may come to control the market if the strategy works, following the approach of claim, demarcate, and control described by Ventresca.[18]

In a difficult market for small companies, the goal has become survival, but the status quo cannot deliver on the goal, because the lack of access to capital has as a knock-on effect, shut off their R&D, and rendered them noncompetitive. Although counter intuitive, the solution for these companies is to embrace more risk not less i.e., lunge forward by acquiring new, externally developed discoveries that could empower a disadvantaged competitor with a superior product or service offering. Basically, adopt a "damn the torpedoes, we're going-in" attitude. However, this requires the alignment of technology availability i.e., patents, access to capital i.e., markets and efficacious technology transfer organizations to complete the cycle, not to mention guts. Invention like potential energy has the capacity to do work but cannot be unleashed unless a successful go-to-market strategy is executed.

U.S. Patents: A Case for Innovation and Job Creation

Patents are the seed corn for innovation for small companies. Specifically, university patents play a disproportionate role here as large corporate inventions are usually captive to the companies that create them and as a result not available for transfer to the emerging companies, with the exception of the open source model. Universities are filing record numbers of patents, which is hugely positive but not sufficient to catalyze the low probability matching process that small companies must go through to have access to an invention that can be leveraged by the SME's business model. For the warehouse of available patents to approach take-off velocity for small companies, the number of issued patents (as opposed to patent applications) needs to increase significantly. The good news is that this is a problem that can be fixed with capital by the U.S. Government with increased funding to the USPTO, for hiring additional patent examiners. The turn-around from patent filing to patent issuance needs to move from 36 months where it currently stands, to something more like a couple of months. The USPTO had approximately 6,242 patent examiners in 2009.[19] The UK Government got the

---

[18] Ventresca (2012).
[19] USPTO (Wikipedia 2012).

memo and recently made proposed modifications to their patent Act of 1977 to among other things, compress the time from filing to issuing a patent in the UK.

As a first level estimate, reducing the backlog of the USPTO would require perhaps a 10× increase in the number of patent examiners. A typical patent examiner earns $70,000–$91,000 per year.[20] Therefore, the patent backlog could be obviated with an increase of roughly $4B per year or roughly 2× times the current USPTO budget.[21]

Increasing the Vitality of the Small Cap Public Marketplace

On April 7, 2012, the Jobs Act was signed into law by President Obama, relaxing Sarbanes–Oxley requirements for small public companies (less than $1B in market cap) for the first 5 years post IPO. Specifically, they won't need to audit their internal controls, although there will be a requirement that they have the internal controls in place. Additionally, the Act allows small private companies to sell shares to up to 499 nonaccredited investors with less than $100 k in yearly income (they can invest up to 5 % of their income). The ACT allows small companies to raise up to $1 M without SEC registration. This remarkable piece of legislation has the potential to unblock equity financing for start-up companies while ushering in a new industry of crowd source financing. Uncharacteristically, both the Senate and the U.S. Congress have supported the ACT, recognizing the connectivity between equity financing, the growth of small companies, and the creation of jobs in the U.S.. However, rebuilding of the small cap ecosystem requires several additional components that have not been addressed by the Jobs Here Act, namely compressing USPTO patent review time lines as described above and, encouraging investments in commercializing new university technologies e.g., technology transfer, through the adoption of reduced corporate taxes on income generated from patented technologies. These so-called patent box strategies have already been implemented in seven countries in Europe (Fig. 2.11) as well as China. The reduction in corporate taxes on income generated from patents and related properties as a result of the patent box adoptions is between 5 and 15 % depending upon the country. PWC, the accounting and business consultancy has articulated a recommendation that the U.S. implements a patent box program similar to those already in place in the UK or Europe[22] and in spite of the valiant efforts of Rep

---

[20] www.inventionstatistics.com

[21] The estimate for the increase in the number of patent examiners needed to reduce USPTO backlog from 36 to 3.6 months, or 10× the current number of examiners, multiplied by an average salary of $70 K per examiner e.g., 10× 6,242×$70,000 = $4.37B p.a.

[22] Peter Merrill et al. (2012) Is it time for the U.S. to consider a patent box? PWC, Tax Notes, March 26th 2012.

Table 1. Comparison of EU Patent Box Regimes and U.K. Proposal							
Tax Factors	**Belgium**	**France**	**Hungary**	**Luxembourg**	**Netherlands**	**Spain**	**U.K.**
Nominal tax rate	6.8%	15%	9.5%	5.76%	5%	15%	10%
Qualified IP	Patents and supplementary patent certificates	Patents, extended patent certificates, patentable inventions, and industrial fabrication processes	Patents, know-how, trademarks, business names, business secrets, and copyrights	Patents, trademarks, designs, domain names, models, and software copyrights	Patented IP or R&D IP	Patents, secret formulas, processes, plans, models, designs, and know-how	Patents, supplementary protection certificates, regulatory data protection, and plant variety rights
Qualified income	Patent income less cost of acquired IP	Royalties net of cost of managing qualified IP	Royalties	Royalties	Net income from qualified IP	Gross patent income	Net income from qualifying IP
Acquired IP?	Yes, if IP is further developed	Yes, subject to specific conditions	Yes	Yes, from non-directly associated companies	Yes, if IP is further self-developed	No	Yes, if further developed and actively managed
Cap on benefit?	Deduction limited to 100% of pretax income	No	Deduction limited to 50% of pretax income	No	No	Yes, six times the costs incurred to develop the IP	No
Includes embedded royalties?	Yes	No	No	Yes	Yes	No	Yes
Includes gain on sale of qualified IP?	No	Yes	Yes	Yes	Yes	No	Yes
Can R&D be performed abroad?	Yes, if qualifying R&D center	Yes	Yes	Yes	Yes for patented IP; strict conditions for R&D IP	Yes, but must be self-developed by the licensor	Yes
Credit for tax withheld on qualified royalty?	Yes	Yes	Yes	Yes	Yes, subject to limitations	Yes, subject to limitations	Yes
Year enacted	2007	2001, 2005, 2010	2003	2008	2007, 2010	2008	2013
Applicable to existing IP?	IP granted or first used on or after Jan. 1, 2007	Yes	Yes	IP developed or acquired after Dec. 31, 2007	Patented IP developed or acquired after Dec. 31, 2006	Yes	Yes

Source: PricewaterhouseCoopers LLP. Information current as of December 31, 2011.

Fig. 2.11 Patent Box Programs in Europe and the U.K. [Merrill et al. (2012)]

Allyson Schwartz or Pennsylvania, who sponsored Bill H.R. 6544 designed to implement a patent box for the U.S., the bill sits in purgatory in the House Ways and Means Committee, dead on arrival. The result is a very tangible disincentive to develop and commercialize patents in the U.S. verses other countries that have adopted the patent box.

Using university technology transfer as an engine room to improve R&D capabilities of small cap companies.

The cost of developing new technologies is beyond the financial and perhaps structural and relationship intellectual capital capabilities of most small cap companies. In the U.S., industrial companies spend about 3.5 % of revenues on R&D, while pharmaceuticals like Merck and Company spend 14 %, computer manufacturing companies 7 %, and biotech's like Allergan as much as 43 % of their revenues on R&D.[23] The direct sponsored research costs to develop a single new technology, from conception to patent filing, at an average the U.S. universities are approximately $4.812 M (AUTM 2010). This estimate should be viewed as ultra conservative, e.g., tip of the iceberg, as it does not include the infrastructure costs such as laboratories, computing facilities, fixed base salaries of researchers, physical facilities, patent expenses, and university technology management expenses among others.[24] Noteworthy, as previously mentioned, approximately 78 %[25] of these patent pending technologies go unlicensed and therefore never become commercialized.

The number of small cap, public companies in the U.S., includes approximately 2,200 firms between $50 M and $300 M in market cap[26] and in the UK there are approximately 1,100 AIM listed small cap companies, of roughly the same size. Few of these companies can afford a $4.18 M new technology development program with a statistical 78 % failure rate. The logical solution is for companies to embrace open innovation and exogenously source technologies from amongst those already developed with taxpayer-funded research at major universities. In spite of its virtue, there are problems with tech transfer for both companies and universities:

1. Most companies still believe that it is better to make rather than buy core technology
2. Most small companies do not have teams that can successfully identify technology acquisition opportunities at leading universities around the world
3. Most small companies are not experienced in licensing technologies from either universities or corporations
4. Most small companies do not have the free cash flow to pay the up-front license fees and subsequent investment in prototypes
5. Most companies are not accustomed to working with tech transfer brokers
6. Most universities do not want to work with brokers, as the traditional tech brokerage model generates revenue for the broker through royalty sharing. The universities already share royalties between the inventor(s) and the institution, and an additional royalty share with a brokerage firm is often difficult to justify

Together, these a priori biases make it difficult to visualize the existing tech transfer brokerage business as being able to effectuate tech transfers systematically. A new model is needed that addresses many if not all of these biases and reduces the friction between university technology suppliers and their customers.

---

[23] Wikipedia (2012).

[24] Based on AUTM data for year 2010, which tabulated sponsored research expenditures at 183 research institutions at $59.1B and number of new patents filed at 12,281 for the year, yielding $59,100,000,000(sponsored research)/12,281 (patent applications) = $4.812 M/patent application.

[25] AUTM (2010).

[26] Capital IQ (2012).

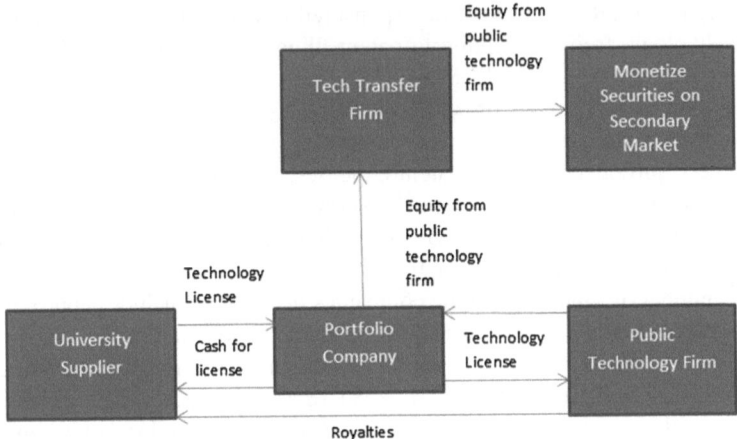

Fig. 2.12 New model for technology transfer 2.0 to reduce friction between technology providers and acquirers thereby facilitating open innovation open innovation (Modified after Gross 2003)

Technology transfer firms, like their university suppliers, are engaged in a technology push enterprise, which is very inefficient.[27] They are intermediaries that seek to create value from ambiguity.[28] A new market-driven technology transfer model, improving upon the work of Gross and Allen (2003) , using knowledge brokering with university networks (Seidel 2012) can be utilized to address this problem. With this model, technologies are sourced for company clients that have a pre-expressed need to a specific technology. Further, this new model needs to incorporate the financing with the technology to empower smaller companies to rapidly begin the commercialization process. To achieve this, these new types of technology transfer firms must have either R&D tax credits, or patent box reduced corporate income tax to mitigate the risk of investing in these new technologies, or they must embrace an equity model, whereby the small public client company utilizes their equity to compensate the tech transfer firm. In the case of the later, it is imperative the tech transfer firm be able to rapidly monetize the equity on the secondary or primary equity markets to ameliorate investment risk (Fig. 2.12). Using this approach, the universities are suppliers of technologies but they do not need to share the royalties with the tech transfer firms, addressing an important bias that the universities have due to their foundational relationships with their faculty inventors.

The weakness of the original Gross (2003) model is that the tech transfer firm creates a portfolio of equity securities in small cap companies which are highly volatile and thinly traded. From an investment perspective this amounts to adverse

---

[27] Geroski (2009).

[28] Ventresca (2012).

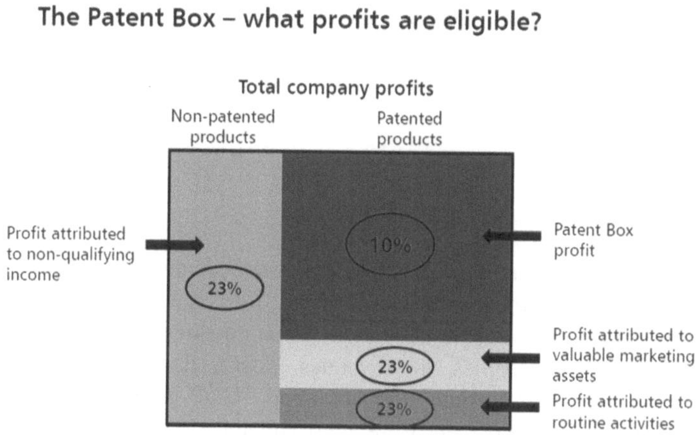

The Patent Box – what profits are eligible?

Fig. 2.13 U.K. Patent Box Tax Relief (HM Treasury, The Patent Box, 10 January 2012)

selection. To address this deficiency, it is necessary to redesign the value chain[29] and effectuate short-term monetization of the equity stakes on primary or secondary markets to ameliorate the risks. Using this new, improved model, technology transfer has been re-visualized to address the broken value chain and facilitate transfers structured as equity based, M&A transactions with additional buyer incentives in the form of R&D tax credits and patent box treatment of income generated from patented technologies (Fig. 2.13).

The following recommendations are put forward as a suggestion to help repair the technology transfer ecosystem, which if implemented may result in significant job creation and GDP growth over subsequent periods in the U.S. and other territories with similar economic fundamentals. The overall mechanism of action would be the creation of marketplace value from university intellectual capital.

Technology

To enhance the pipeline of investable technologies which would encourage small cap private and public company investment, immediately increase the funding of the USPTO to enable them to increase the number of patent examiners on staff to reduce patent application review time from 36 months to a couple of months. Preliminary estimates are that this would cost $4B. This compressed review period would set a new standard for the world, create a competitive advantage for the USPTO and the U.S. small cap market, and enhance the ability of small firms to attract investment capital. It is likely that approximately 2.5 million new jobs

---

[29] Ventresca (2012).

would be created as a result of the accelerated review process. The cost p.a. would be approximately $1,760 per job created. As a point of comparison, the Recovery Act of 2008 spent $240,000 for each job created.[30]

Market

To address the small cap market failure and facilitate an increase in both IPO's and follow-on financings, immediately follow the Jobs Act of 2012 with further reduced regulatory requirements for small cap companies such as suspension of Sarbanes–Oxley for companies with revenues under $1B. Data indicate that more than 500 IPO's per year are necessary to support GDP growth of 3 % and continued job growth. In 2011, there were approximately 150 IPO's in the U.S..

Organization

To neutralize the inability of small companies (public and private) to fund R&D, the U.S. Congress should extend the R&D tax credit, which expired in December 2011, to encourage technology venture investment by firms. Specifically, the development of patent box incentives aimed at companies investing in university technologies, would be hugely beneficial for encouraging tech transfer as both an industry and a tool for enhancing the success of small companies. This, combined with a newly developed model of tech transfer outlined here, would create reinforcing linkages between patent issuances, the capital markets, and the growth of small companies.

Alternatively, preferential corporate tax treatment (e.g., patent box programs similar to those in place in Europe, the UK, and China) should be offered to companies that commercialize patents, which simultaneously encourages innovation and corporate venturing (investment) with new technologies.

Nine countries including China, Belgium, France, Ireland, Luxemburg, Netherlands, Spain, Switzerland, and the U K have recently passed a patent-related tax shield called the "Patent Box," in reference to the box checked on the tax return. The goal of all of these programs is to encourage the commercialization of innovations as distinct from R&D tax credits which encourage the investment in new discoveries. In the UK, for example, under the new patent box tax program, corporations that commercialize patented technologies can avail themselves of a preferential 10 % corporate tax rate (vs. 24 %) on profits generated from these technologies.

---

[30] Luhby (2009).

Taken together, these recommendations would enhance competitiveness and strengthen GDP in any country that adopts them by increasing the success rate of its most important corporate constituency small companies. The Jobs Act in the U.S. is an excellent first step, but it needs to be implemented concomitant with the USPTO reform Act and a tech transfer patent box to reduce the corporate tax rate to catalyze small company growth and incentivize corporate commercialization of new innovations.

References

AUTM U.S. Licensing activity survey highlights: FY. 2010. http://www.autm.net/AM/Template.cfm?Section=FY_2010_Licensing_Survey&Template=/CM/ContentDisplay.cfm&ContentID=6874. Accessed 29 Jan 2012.

Borer J. Personal communication, Rodman and Renshaw. 2012. New York, Jan 2012.

Capital IQ. https://www.capitaliq.com/home.aspx (2012). Accessed 6 Mar 2012.

Chesbrough H. Open innovation: the new imperative for creating and profiting from technology. Boston: HBS Press; 2003.

Geroski P. The evolution of new markets. Oxford: Oxford University Press; 2009.

Granovetter M. The strength of weak ties. Am J Sociol. 1973;78:1360–80.

Gross C, Allen J. Technology transfer for entrepreneurs. Westport: Praeger Publishers; 2003.

Gross C. U2B: A new model for technology transfer. BioEntrepreneur. 2003;21.

Image of Porter's Five Forces, from Investopedia.com. 2012. http://www.investopedia.com/features/industryhandbook/porter.asp#ixzz1qi2irqCX. Accessed 31 Mar 2012.

Inventionstatistics.com. http://www.inventionstatistics.com/Patent_Office.html (2010). Accessed 8 Apr 2012.

Levensohn P. http://www.pascalsview.com/pascalsview/2010/08/connecting-the-dots-how-new-job-creation-ipo%E2%80%99s-and-venture-capital-in-america-are-intimately-linked.html (2010). Accessed 4 Mar 2012.

Luhby T. Stimulus creates 640,000 jobs, CNNMoney, Special Report: The Rescue. 2009. http://money.cnn.com/2009/10/30/news/economy/Stimulus_jobs_created/index.htm.

Mishkin F. Monetary Policy and the Dual Mandate, Speech given at Bridgewater College, Bridgewater, Virginia, 10 Apr 2007. http://www.federalreserve.gov/newsevents/speech/mishkin20070410a.htm. Accessed 26 Mar 2012.

Nothaft H, Michel P. New York Times Opinion article, Inventing Our Way Out of Joblessness 5 Aug 2010.

Obama B. Forum on modernizing government, white house, 14 Jan 2010. http://www.inventionstatistics.com/Patent_Backlog_Patent_Office_Backlog.html. Accessed 30 Mar 2012.

Merrill P, Shanahan J, Gomez J, Glon G, Grocott P, Lamers A, MacDougall D, Macovei A, Montredon R, Vanwelkenhuyzan T, Cernat A, Merriman S, Moore R, Muresan G, Van Den Berghe P, Linczer A. Is it time for the United States to consider the patent box? Tax notes. LLC: PricewaterhouseCoopers; 2012.

Porter ME. How competitive forces shape strategy. 1979. Harvard Business Review, March/April 1979.

Perry M. U.S. patent activity continues to grow, Blog for Economics and Finance. 28 Nov 2010. http://mjperry.blogspot.com/2010/11/american-exceptionalism-continues-to.html. Accessed 4 Mar 2012.

Quinn G. IPWatchdog, Inc., Posted: 2 Jan 2011. http://www.ipwatchdog.com/2011/01/02/why-patents-matter-job-creation-economic-growth/id=14170/. Accessed 3 Mar 2012.

Ritholtz B. The big picture. 30th Jan 2009. 11:15AM http://www.ritholtz.com/blog/2009/01/historical-chart-of-initial-public-offerings/. Accessed 1 Mar 2012.

Schramm C. Jump-starting job creation, CNBC, Wed 02 Dec 09 | 02:22 PM ET, The best path to tackling double-digit unemployment, Kauffman Foundation. 2012. http://video.cnbc.com/gallery/?video=1348493141. Accessed 24 Mar 2012.

Seidel V. Handout for strategy and innovation course, models of innovation in technology based markets, EMBA8. Said Business School, Oxford: Oxford University. Jan 2012. p. 1, slides 4, 6.

Serchuk D. Decimalization and its discontents, Forbes.com. 10 Mar 2009. http://www.forbes.com/2009/03/09/decimalization-uptick-rule-intelligent-investing-volatility.html. Accessed 3 Mar 2012.

The Tech-Online Edition, MIT, Volume 130, Issue 23: Friday, 30 Apr 2010, Accessed 3 Mar 2012.

University of Toledo. 2012. http://www.eng.utoledo.edu/eecs/publications_patents/commercial.html. Accessed 29 Jan 2012.

Utterback JM. Invasion of a Stable Business by Radical Innovation: Natural Ice Industry. Mastering the Dynamics of Innovation, Chapter 7, HBS Press, Boston, MA, 1994.

U.S. Patent and Trademark Office (Wikipedia, 2012). http://en.wikipedia.org/wiki/USPTO, Accessed 8 Apr 2012.

Ventresca M. Handout #4, p. 20 & Handout #6, p. 7, for Strategy and Innovation course, EMBA8. Said Business School, Oxford University, January 2012. p 20.

Weild D, Kim E. Market structure is causing the IPO crisis and more, Grant Thornton, LLP, 2010.

Weitekamp R, Pruitt B. Job growth in U.S. driven entirely by startups, kauffman foundation study. 2010. http://www.kauffman.org/newsroom/u-s-job-growth-driven-entirely-by-startups.aspx. 7 July 2010.

World Intellectual Property Indicators. 2012. Edition, World Intellectual Property Organization, Geneva.

Chapter 3
Intellectual Capital: The World's Fastest Growing Asset Class

Economic growth and competitiveness is highly dependent on productivity. One of the most commonly used measures of productivity is Gross Domestic Product (GDP) per hour worked.[1] According to the US Bureau of Labor Statistics, the US ranked third in labor productivity worldwide in 2011[2] (Fig. 3.1) and similarly in GDP per capita[3] (Fig. 3.2), a basic indicator of a country's economic vitality.

Yet, in spite of the good ranking of the US regarding GDP per capita, the real economy is quite a distance from the robustness of the 1990s. Currently, 75 % of Americans have savings equal to less than 6 months of their living expenses.[4] More than 46 million Americans live in poverty,[5] which amounts to one in six individuals. The Food Stamp program, an updated version of the soup kitchens of the depression era, serves 25 % of American children.[6] At the time of this writing in early 2013, 12.3 million Americans (7.9 %) were unemployed[7] and an additional 9 % were underemployed (working part time as opposed to their goal to work full-time). Also, nearly seven million Americans have given up looking for work[8] and are not counted in the reported unemployment statistics. *The new normal*[9] demonstrates the unfortunate characteristic of a reduced standard of living for the majority of the citizens of the OECD. The solution, however, is surprisingly

---

An erratum to this chapter is available at 10.1007/978-3-319-00281-1_8

---

[1] OECD (2008), 11.
[2] US Bureau of Labor Statistics, Division of International Labor Comparisons. http://www.bls.gov/ilc/intl_gdp_capita_gdp_hour.htm#chart03
[3] IBID.
[4] http://www.cbsnews.com/8301-500395_162-57459596/survey-half-of-americans-have-insufficient-emergency-savings-quarter-have-none-at-all/.
[5] http://www.povertyusa.org/
[6] http://www.presstv.ir/detail/2013/01/11/283026/one-in-four-us-kids-on-food-stamps/
[7] http://www.bls.gov/news.release/empsit.nr0.htm
[8] http://www.csmonitor.com/USA/Politics/2012/0908/Why-have-millions-of-Americans-given-up-looking-for-work
[9] http://www.bloomberg.com/news/2011-10-03/pimco-s-gross-says-global-recession-risk-is-overtaking-new-normal-forecast.html

C. M. Gross, *Too Good To Fail*, Management for Professionals,
DOI: 10.1007/978-3-319-00281-1_3,
© Springer International Publishing Switzerland 2013

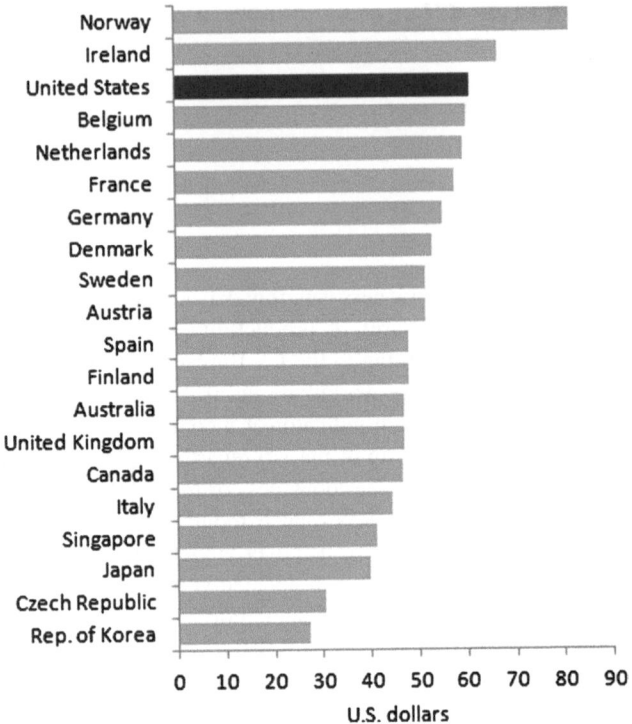

GDP per hour worked, 2011
Converted to U.S. dollars using 2011 PPPs

Fig. 3.1 *GDP* per hour worked in 2011

simple, namely, enhance productivity. How to achieve this, however, is a bit tricky. It requires immediately putting into play our trump suit; namely, commercializing the pent-up and growing number of discoveries of the world's research universities, to create marketplace value and improve the quality of life, through accelerated corporate growth.

Another more complex measure of productivity is total factor productivity (TFP). TFP is the contribution of technology and innovation to economic growth. Intellectual capital is the world's fastest growing asset class, which on a country-by-country basis is highly dependent on TFP. TFP may be deconstructed to include educational preparation of the labor pool, technology developed and in use (i.e., patents), educational institutions, government institutions, and policies which effect the production of innovation from these various tributaries of intellectual capital. Specifically, TFP includes those factors not included in the production function used to calculate GDP.[10] As a result, TFP may be thought of as a residual

---

[10] Miles and Scott (2005), p. 49.

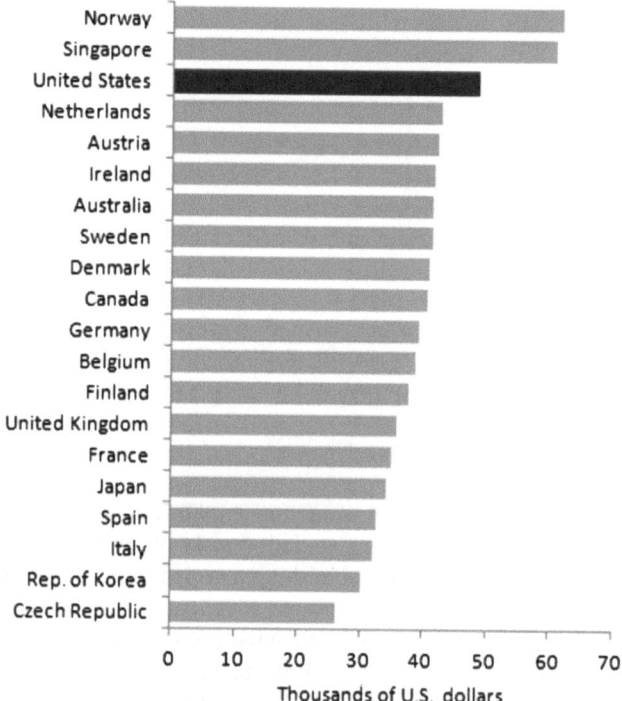

GDP per capita, 2011
Converted to U.S. dollars using 2011 PPPs

Fig. 3.2 *GDP* per capita in 2011. *Source*: U.S. Bureau of Labor Statistics, Division of International Labor Comparisons

(Solow residual) that explains GDP minus capital stock and hours worked. TFP may be calculated with the Cobb–Douglas method[11] as follows:

$$Y = A \times K^{\alpha} \times L^{\beta}$$

where:
Y = total output (real GNP)
A = TFP
K = capital input
L = labor input
α = returns to capital (between 0 and 1) (approximately 0.33)
$\beta = 1 - \alpha$ = returns to labor (approximately 0.67)

---

[11] Miles and Scott (2005), p. 49 and Wikipedia (2012).

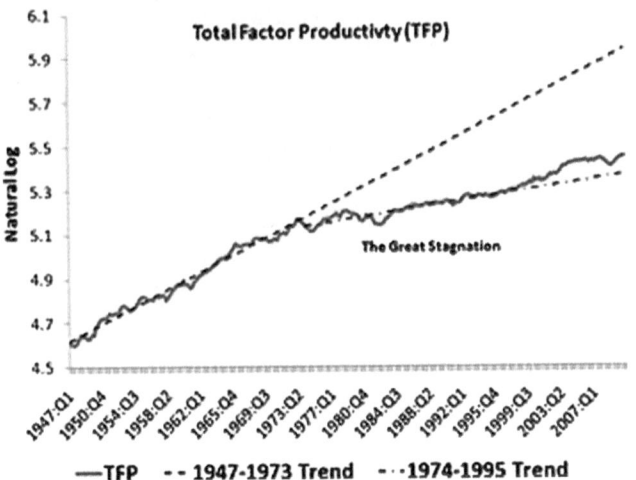

Fig. 3.3 Total factor productivity (Beckworth 2011)

TFP plays an important role in economic growth (Comin 2008). Long-term growth is dependent upon capital accumulation and improvements in TFP (Miles and Scott 2005). In the face of the diminishing marginal product of capital (MPK) for wealthy nations such as the US, growth is reliant on improvements in technological progress as well as education of the workforce. From 1948 to 1973 in the US, growth was all about TFP, followed by stagnation and a slight boost higher around 1990 (Beckworth 2011) (Fig. 3.3). Higher TFP enables both higher output per hours worked and a higher steady state of capital stock. Both of these factors contribute to greater GDP per capita. TFP also includes the effect of government policies e.g., availability of patent protection, legal institutions e.g., enforcement of contracts regarding property rights and importantly, functioning capital markets to provide the growth capital for commercializing innovations that can lead to enhanced productivity, such as on-line commerce, new materials and microprocessors, and to name a few. "GDP is a measure of the value added that a society produces."[12] The valued add is created and maintained by entrepreneurial activities versus consuming value added through rent seeking activities. Rent seeking serves as a drag on GDP and lowers TFP. For the invisible hand to work its magic in a positive manner, the rule of law must prevail and manifest itself in sensible corporate governance and policies that promote civil liberties, individual freedom, and risk taking e.g., investment in innovation and new enterprise formation.

According to Law (2000), "TFP of an economy only increases if people work smarter and learn to obtain more output from a given supply of inputs." Although there has been a continuous decline in TFP growth rates in the US since 1948, there was an interesting reversal of this trend in 1990–2001 (Fig. 3.4).

---

[12] Miles and Scott (2005), p. 96.

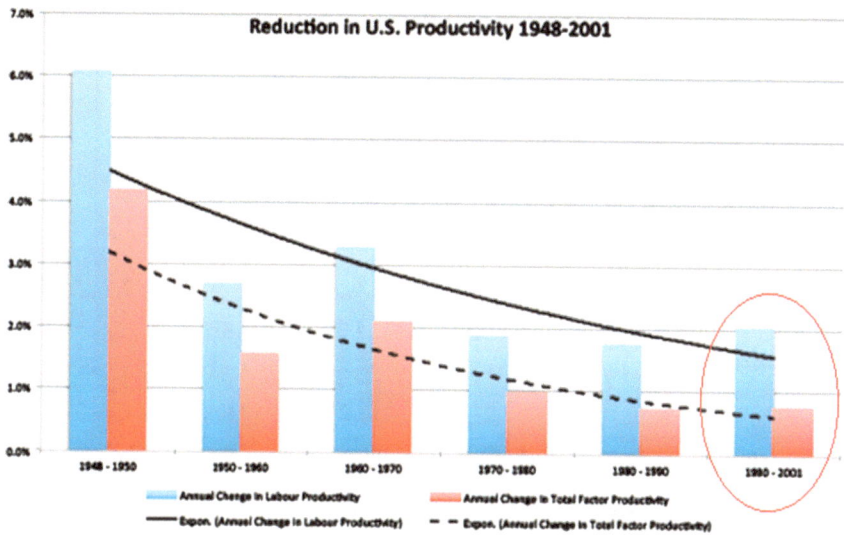

Fig. 3.4 US productivity and TFP growth rates have declined between 1948 and 2001 (Calculated data from The US Bureau of Labor Statistics, http://www.bls.gov/news.release/prod3.t04.htm)

US Patents as the Source of Innovation and Job Creation

For the last 20 years, the number of US patent applications has risen dramatically (Fig. 3.5) from approximately 200,000 in 1990 to > 550,000 in 2001. Interestingly, patent issuances have not kept pace, with about 180,000 issued p.a. in 1990 to about 330,000 p.a. in 2001.

In 2006, for example, more US patents were granted than in any previous year. The 1,800,000 + patents granted in the 10-year period between 1998 and 2008 were greater than the all of the patents issued in the 25-year period of 1963–1987.[13]

Clearly, the patent invention engine room is moving forward at a record clip; however, innovation is more than just the number of patents issued, it also requires the commercialization of the invention i.e., capital, to result in productivity improvements.

A closer look at the 1990–2001 periods (Fig. 3.6) reveals a particularly large run-up in patent activity preceding the dot com reversal. This is consistent with the large inflows of capital investment into technology start-ups during the same period. No causality is implied here, as a bullish market creates its own virtue; however, the availability of investment returns applied to intellectual property can potentially or at least theoretically have the knock-on effect of increasing TFP.

---

[13] Perry (2010).

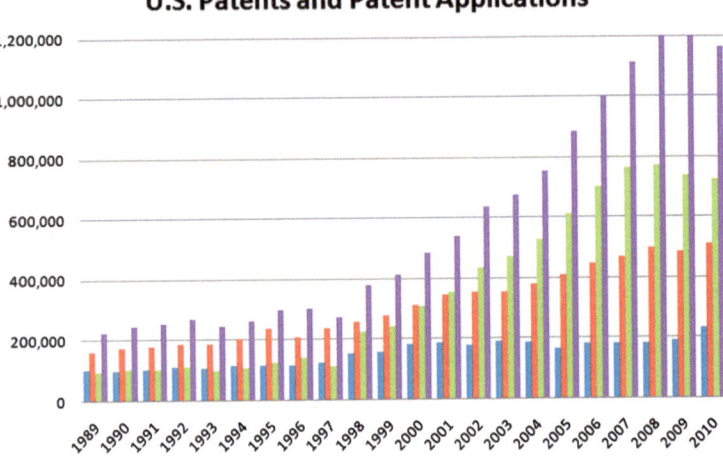

Fig. 3.5 US patents and patent application spatent applications (From Perry 2010 using USPTO data)

Fig. 3.6 Growth of patent activity between 1990 and 2001 (Schoen 2004 data from USPTO)

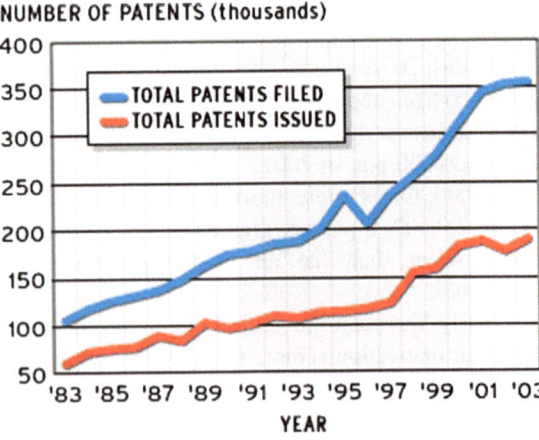

After 2000, the divergence between patents filed and patents issued increased dramatically (Fig. 3.7). By 2010, of the 475,000 utility patent applications filed only about 200,000 were issued. There may be a number of reasons for this, including a reduction of funding for the US Patent and Trademark Office.[14]

---

[14] Nothaft and Michel (2010).

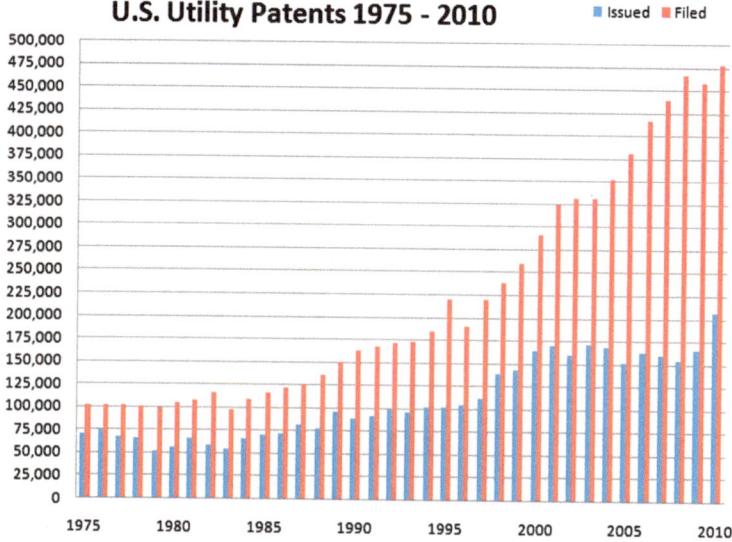

Fig. 3.7 The divergence between patents filed and issued increases (Quinn 2011)

It has been estimated that there were 1.2 million patent applications currently awaiting examination by the USPTO[15] and that for every patent issued between three and ten new jobs are created. If the USPTO backlog were resolved, the authors believed that approximately 2.5 M new jobs would be added to the business sector in the US.

Small Cap Market Failure in the US IPO Market

For TFP to increase, technological innovation is paramount. Patents are a good proxy for innovation but not the thing itself, as innovation is a knock-on effect of invention plus the capital market activity necessary for the commercialization of these improved products and processes. Beginning in 1990, the number of IPOs increased dramatically as investors were interested in deploying capital with early stage, technology enriched firms (Fig. 3.8). A period of job creation, economic surpluses, and productivity improvements resulted. However, excesses occurred as well. The bitter taste of the ENRON and WorldCom bankruptcies resulted in a public outcry for reform, and as result a series of financial reform acts were implemented with the intent to reduce the occurrence of future fraudulent activities. Research and investment banking were separated, analysts could no longer get individual corporate briefings (Regulation FD), transparency rules, and improved

---

[15] Nothaft and Michel (2010).

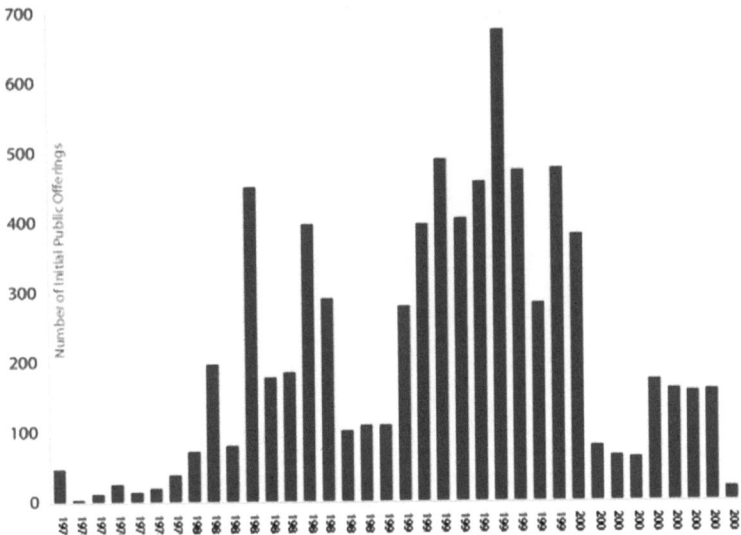

Fig. 3.8 US IPOs 1973–2008 (Ritholtz 2009, http://www.ritholtz.com/blog/2009/01/historical-chart-of-initial-public-offerings/)

verification of controls were implemented (Sarbanes–Oxley et al.,). Additionally, technological improvements created the industry of on-line trading, beginning the long disintermediation of stock brokering activities. Further, decimalization of quotes on the stock exchanges in 2001 reduced the quoted price on stocks from 6.25 cent intervals (1/16 of a dollar) to one cent. This modernization had the anticipated effect of reducing bid-ask spreads and the unintended effect of dis-enfranchising market makers, as narrower spreads meant smaller profits and less profit means that market makers have less capital to invest. The result was a reduction in liquidity across the board (Serchuk 2009).

All of these factors had the makings of a perfect storm for the capital markets in 2000 and beyond. Together, in short, they made it difficult for small cap investment banks to make a living. As a result, most went out of business and the number of IPOs fell off dramatically after 2000. Currently, there are only about a dozen small cap investment banks in the US, whereas 15 years ago there were hundreds.

With startling efficiency, the law of unintended consequences combined with a reduced public appetite for equities had swiftly decimated the public equity markets for small cap stocks from 2001 onward.

Prior to 2000, 80 % of IPOs raised less than $50 M [Levensohn (2010)], afterwards, only 10 % would raise < $50 M. As a telling example, "Intel Corporation went public in 1971 with an $8 million IPO and a mere $53 million valuation," according to Stewart (2010).[16] Currently, few companies can raise < $50 M in an IPO.

---

[16] Why $10 M IPO's matter, http://www.urgentspeed.com, 2010.

In addition to the devastation of the small cap IPO market following 2000, the market for follow-on offerings of public companies that are seeking additional capital raises has also evaporated. IPOs are essential for raising capital to commercialize new discoveries as well as for job and GDP growth. In 2011, there were 134 IPOs in the US, with an average size of $265 M (The Business Review 2012), far below the demarcation described by Weild and Kim.[17]

IPOs and Job Creation

Eric Cantor (2011), a six-term US congressman from Virginia, said approximately 70 % of all new job creation in the US is from small companies.[18] Small companies are the engine room for economic growth in the US. According to Pascal Levensohn, "If the total annual number of Initial Public Offerings (IPO's) in the US does not exceed 500, which studies show is the level required to support 3 % annual US GDP growth, the US will not generate the job growth necessary to rekindle meaningful sustainable GDP growth in the US."

According to the Kauffman Foundation, a recognized authority on entrepreneurship, start-ups are responsible for virtually all new job creation in the US since 1977. Nationwide, new firms add an average of three million jobs in their first year while established organizations loose roughly one million jobs per year.[19] Further, 92 % of the job growth in companies that go public, occur post IPO (Fig. 3.9), indicating the importance of functioning capital markets, especially for small firms, if technological innovation is to translate into enhanced TFP, growth in employment, and GDP.

Commercializing Research

For inventions to create economic value they must be commercialized. There are patents and there are blockbuster patents. University discoveries are often in a class of their own, and this intellectual property has produced some of the most successful companies and products in the world.

Technology transfer is necessary to create a robust pipeline of technology acquisition opportunities for small firms. This open innovation gestalt enables smaller firms to neutralize their R&D weakness (as compared with their better financed brethren) by exogenously acquiring new discoveries to strengthen their product/service offerings. Universities and Federal Research Laboratories (i.e.,

---

[17] Weild and Kim (2010).

[18] Quoting from US Bureau of Labor Statistics data.

[19] Weitekamp and Pruitt (2010).

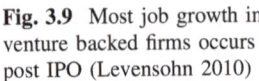

Fig. 3.9 Most job growth in venture backed firms occurs post IPO (Levensohn 2010)

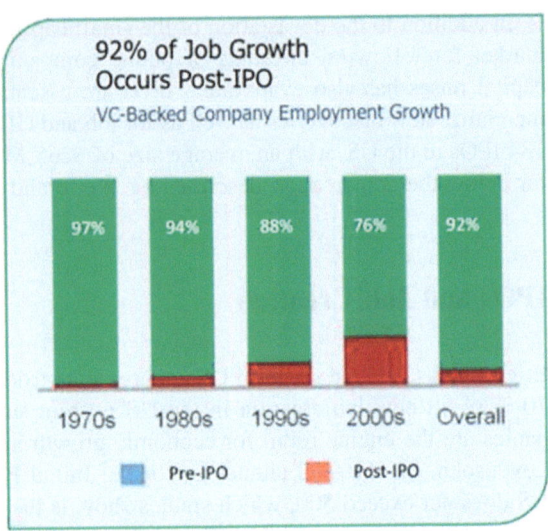

NASA, Los Alamos, Dept. of Energy, etc.) are a likely source for these discoveries as corporate R&D in larger firms, usually and logically remains captive to the firm that finances it.

Comin and Hobijn (2010) have demonstrated that technology transfers from the US were an important element in the post-World War II recovery of Europe and Japan, accelerating their TFP beyond what it was pre-war, allowing them to produce significant post-war economic performance (GDP growth) and not just catch-up economics or recidivism. What worked for Europe and Japan after 1948 may well be helpful at invigorating TFP in the US and Europe in 2013 and beyond.

TFP and Patent Applications Between 1948 and 2009

To analyze the association, if any, between US TFP data and US patent data, two data sets have been merged; US TFP data were acquired from the San Francisco Federal Reserve Bank for the years 1948–2009 and patent data (both applications filed and patents issued) were acquired from US Patent and Trademark Office for the years 1948–2009.

The change in TFP rates between 1948 and 2009 shows a downward trend (Fig. 3.10) consistent with the US BLS data. However, the number of patents issued in the US has soared from 1945 to 2009. Yet, as previously mentioned, severe patent office backlogs have significantly reduced the number of patent issuances.

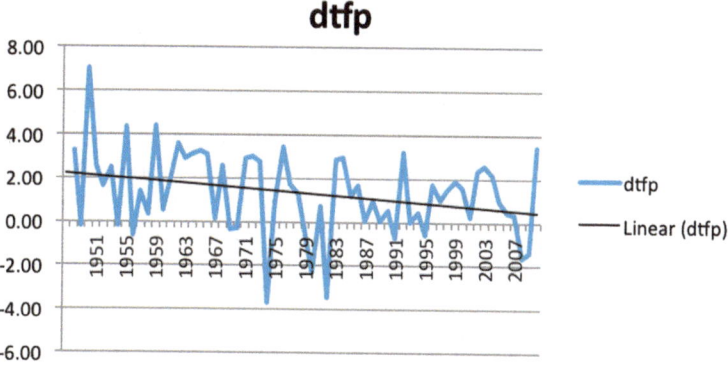

Fig. 3.10 Changes in TFP in the US between 1948 and 2009

Initial analyses of the data between patent applications and TFP have yielded no obvious trends or relationships. This may be due to a number of factors including that technology diffusion rates i.e., it may take an issued patent several years before it creates marketplace value and is dependent upon capital availability for commercialization (i.e., venture capital investment, the number of IPOs, follow-on offerings, and bank lending policies) all of which vary annually. The data did not take into account these factors.

Others have also documented the weak relationship between patent stock and TFP.[20] They have offered that patents may be a rather weak proxy for knowledge, and knowledge is the major determinant for TFP growth and that with or without patents, R&D has a minimal effect on TFP. They concluded that the effect of knowledge on TFP is complex and subject to the diffusion of knowledge over time, and therefore difficult to measure. Additionally, Comin and Mulani (2005) also have reported an imprecise relationship between R&D and TFP at the macro level. They offered that general innovation (as distinct from specific patented improvements, i.e., technology spill overs) such as mass production or the Internet, which on balance and writ large are not patentable, are responsible for the growth of TFP and concomitant productivity gains. An additional factor that might contribute to the weak relationship between R&D and TFP might be the contribution from technology commercialization, which was not factored into any of the aforementioned studies, yet is necessary for creating value from R&D and its patent fruits.

There are of course many other possible explanations for the weak observed relationship between patents and TFP. Specifically, as mentioned, it would be interesting to deconstruct the patent data into those inventions that are actually commercialized. It is logical or at least plausible that such a reduced and rarified data set may well serve as a better predictor of TFP.

Although I have found that no clear pattern has emerged between patent data and TFP in the last 60 years, even though sensibility would anticipate some level

---

[20] Abdih and Joutz (2005).

of correlation. However, Chu (2007) has concluded that "35 to 45 % of the long-run TFP growth in the US is driven by R&D," for which in his research, patents were a proxy. Additionally, Zachariadis (2002) has determined that, "R&D intensity has a positive impact on the rate of patenting, the rate of patenting has a positive effect on technological progress, and, finally, technological progress has a one-to-one relation with the growth rate of output per worker. Moreover, the intensity of aggregate manufacturing R&D is shown to have a stronger impact on the rate of patenting than own-industry R&D. This implies technological spillovers across manufacturing industries." This is plausible but far from proven.

More to the point, for the US to support 3 % GDP growth, approximately 500 IPOs p.a. is needed.[21] New company formation drives job growth, post-IPO companies drive job growth; patent issuances drive job growth and job growth produces GDP growth.

It is logical, that increased patent issuances through improved USPTO efficiency, combined with more readily available capital or preferential tax incentives (e.g., patent box regimes) (Merrill et al. 2012) to incentivize their commercialization, are likely to enhance TFP. Due to the current complex and burdensome regulatory environment in the US, this will necessitate the resuscitation of the small cap public equity markets as well as the enactment of patent box tax incentives. Enhancement of TFP, job and GDP growth in the US may well depend on it.

References

Abdih Y, Joutz F. Relating the knowledge production function to total factor productivity: An endogenous growth puzzle. IMF Working Paper. pp. 1–40, http://ssrn.com/abstract=888120, http://home.gwu.edu/~bmark/Abdih-Joutz-Text-Flood.pdf (2005). Accessed 5 Mar 2012.

Beckworth D. The great stagnation and total factor productivity. http://macromarketmusings.blogspot.com/2011/02/great-stagnation-and-total-factor.html (2011). Accessed 12 Jan 2012. Constructed from data from John Fernald (San Francisco Federal Reserve Bank).

Cantor E. http://www.politifact.com/virginia/statements/2011/dec/30/eric-cantor/cantor-says-small-businesses-create-70-percent-us (2011). Accessed 29 Jan 2012.

Chu A. Economic growth and patent policy: Quantifying the effects of patent length on R&D and consumption. University of Michigan, October 2007, MPRA Paper No. 5476, posted 07. November (2007). Accessed 6 Mar 2012.

Comin D, Hobijn B. Technology diffusion and post-war growth, HBS, working paper 11–027, 2010.

Comin D. Total factor productivity. 2nd ed. In: Durlauf S, Blume L, editor. The new palgrave dictionary of economics. Basingstoke: Palgrave Macmillan; 2008.

Comin, D, Mulani S. A theory of growth and volatility at the aggregate and firm level. National Bureau of Economic Research working paper No. 11503, 2005.

Law, M. T., Productivity and Economic Performance: An overview of the Issues, Public Policy Sources, No. 37. 2000. Accessed 12 Jan 2012. http://www.afroarticles.com/article-dashboard/Article/The-Global-Productivity-Trends-and-the-Changing-World-Economic-Order/199319 Accessed 3 Mar 2012.

---

[21] According to Levensohn (2010).

Levensohn P. http://www.pascalsview.com/pascalsview/2010/08/connecting-the-dots-how-new-job-creation-ipo%E2%80%99s-and-venture-capital-in-america-are-intimately-linked.html (2010). Accessed 4 Mar 2012.

Merrill P, Shanahan J, Gomez J, Glon G, Grocott P, Lamers A, MacDougall D, Macovei A, Montredon R, Vanwelkenhuyzan T, Cernat A, Merriman S, Moore R, Muresan G, Van Den Berghe P, Linczer A. Is it time for the United States to Consider the Patent Box? Tax Notes, 26 Mar 2012, PricewaterhouseCoopers, LLC. 2012.

Miles D, Scott A. Macroeconomics: Understanding the wealth of nations. England: John Wiley and Sons; 2005.

Nothaft H, Michel P. New York Times Opinion article. Inventing Our Way out of Joblessness 5 Aug 2010.

OECD. Compendium of Productivity Indicators. OECD. 2008.

Perry M, U.S. patent activity continues to grow, Carpe Diem Blog, 28 Nov 2010 http://mjperry.blogspot.com/2010/11/american-exceptionalism-continues-to.html (2010). Accessed 4 Mar 2012.

Quinn E, IPWatchdog, Inc., Posted: Jan 2, 2011, http://www.ipwatchdog.com/2011/01/02/why-patents-matter-job-creation-economic-growth/id=14170/ (2011). Accessed 3 Mar 2012.

Ritholtz, B. The Big Picture, January 30th, 2009, http://www.ritholtz.com/blog/2009/01/historical-chart-of-initial-public-offerings/ (2009). Accessed 1 Mar 2012.

Schoen J. msnbc.com, updated 4/27/2004, http://www.msnbc.msn.com/id/4788834/ns/technology_and_science-tech_and_gadgets/t/us-patent-office-swamped-backlog/ (2004). Accessed 3 Mar 2012.

Serchuk D. Decimalization and Its Discontents, Forbes.com, March 10, 2009, http://www.forbes.com/2009/03/09/decimalization-uptick-rule-intelligent-investing-volatility.html (2009). Accessed 3 Mar 2012.

Stewart J. Why $10M IPO's matter, http://www.urgentspeed.com, 2010

The Business Review, March 6th, 2012 (quoting PWC, LLC), http://www.bizjournals.com/albany/morning_call/2012/03/ipos-were-down-20-percent-in-2011.html.

Total factor productivity, Wikipedia. http://en.wikipedia.org/wiki/Total_factor_productivity. Accessed 1 Mar 2012.

Weild D, Kim E. Market structure is causing the IPO crisis and more, Grant Thornton, LLP, 2010.

Weitekamp R, Pruitt B. Job Growth in U.S. Driven Entirely by Startups, Kauffman Foundation Study, http://www.kauffman.org/newsroom/u-s-job-growth-driven-entirely-by-startups.aspx, 7 July 2010.

Zachariadis M. R&D, Innovation, and Technological Progress: A test of the Schumpeterian Framework without Scale Effects, Department of Economics, 2107 CEBA, Louisiana State University, Baton Rouge, LA, 2002. http://www.bus.lsu.edu/economics/papers/pap02_18.pdf, Accessed 5 Mar 2012.

Chapter 4
Network Strategies for Growing Emerging Markets

The implementation of the appropriate network strategy is vital for the development of new and immature markets such as once typified by biotech, Internet commerce, and hybrid powered vehicles. A sagacious course of action can rapidly accelerate market development and enhance the quality of participant relationships and therefore the potential for revenue generation by companies. There are many network strategies and amalgams that can be leveraged, including[1]:

Primordial: Similar nodes, common social identity, i.e., diamond merchants.

Supply Chain: Dissimilar nodes, common company, or work identity, i.e., a car company's suppliers.

Invisible College: Similar and dissimilar nodes, common interest in data, knowledge, or innovation, i.e., universities research alliances.

Strategic: Similar and dissimilar nodes, different interests but goals concatenate (or at least are co-planer), i.e., Silicon Valley VC marketplace and eco-system.

Once the network strategy is selected, market ownership is facilitated by implementing a staged-like process with the claim-demarcate-control framework.[2] These stages consist of:

Claiming a market: Defines a position developed by experience, reason, or analogy and describes how the new approach might be superior. It is important to establish the credentials of the source of the market lens, giving it authority and potential energy for capital commitment for the journey.

Demarcating a market: Develops and executes a business model that embraces the market boundaries and creates value in addition to what can be had elsewhere, for each of the participants (new enterprise, customers, suppliers, and when applicable, regulators). This necessitates the forming of strategic alliances and win–win agreements between a company and its suppliers and customers. These agreements must deliver profit sharing among participants (whether for-profit or eleemosynary entities).

Controlling a market: This is the build-out stage, usually defined by increasing network affiliations, growth of revenues, and the emergence of the company as a

---

[1] Ventresca (2012).
[2] IBID.

C. M. Gross, *Too Good To Fail*, Management for Professionals,
DOI: 10.1007/978-3-319-00281-1_4,
© Springer International Publishing Switzerland 2013

dominant player. Activities here reinforce growth with acquisitions, standard setting, significant investment in branding, and public relations, with the modus operandi of holding one's self out as the leader in the industry, as Jobs did with Apple in the early years of the PC industry and Elon Musk has done with Tesla, in the electric powered vehicle segment.

Networks may be characterized as having, strong and weak ties, structural holes, and may be made more efficient through the addition (e.g., technology transfer) or subtraction (e.g., stock brokers) of intermediaries. Brokers are Intermediaries who seek to create value by facilitating transactions through their experience and the added knowledge they may provide. Technological advances have shown this to be a time limited phenomena.

Structural holes in networks describe a gap in connectivity in the customer, supplier, and company networks. Strong ties are "a combination of the amount of time, the emotional intensity, the intimacy (mutual confiding), and reciprocal services"[3] invested. Weak ties are informal connections that facilitate unstructured, pragmatic connectivity.

The cost of maintaining strong network nodes may not be worth-the-carriage over time and will give way to weak ties which are more allowing, less confining, and costly, and in many ways can be more opportunistic.[4] As an example, the migration of IBM into the PC business, where suppliers had strong ties with the firm, to one where a supplier (Lenovo Group) ends up acquiring the PC business from IBM in 2005. In this case the onetime strong nodal supplier now has a weak node relationship. With IBM serving as a potential referral source for Lenovo, for IBM consulting clients who may want to acquire PCs. Weak ties do not imply a pejorative, as they may be preferred under certain conditions and may provide a competitive advantage over strong ties in achieving corporate goals. This is especially true in the global university discovery network, where new technologies can appear rapidly from a myriad of research institutions and often need to be pieced together to create a workable solution.

A Brief Example from Technology Transfer

A priori one may view the technology transfer service industry through the lens of the invisible college network as it relies on universities supplying a pipeline of new discoveries to the commercial sector. Intermediaries such as technology transfer firms usually have strong ties with a wide range of universities and intellectual property experts and weak ties with venture capital providers and commercial clients who are targets for the intellectual property. It is characterized as a value-added strategy that combines go-to-market expertise with the IP. In this

---

[3] Granovetter (1973).

[4] Peng and Zhou (2005).

traditional model, most tech transfer firms assist the developers of the technology, in a brokerage capacity, to bring their innovations to market (technology push).

In a new model introduced by Gross,[5] he describes a market-driven technology transfer model that begins with the end client (the corporate customer) who has a strong tie with the technology transfer firm. This is combined with a large array of university alliances characterized by weak ties that serve to supply first information on technologies available for license, to be followed by the transfer of specific, corporate client approved technologies, i.e., hence market-driven or technology pull. With technologies it is more efficient to pull, then push. This also addresses holes in the university network as they normally do not have strong ties with a large range of firms, which renders them demand constrained and long on supply. This new model for technology transfer is essentially a morphing from a collegiate network model to a strategic network model. A company that previously utilized the new model[6] and a version of the CDC framework became the largest technology transfer firm in the U.S. in 2006, transferring a new technology roughly every two weeks, with sales exceeding $50 million in that year.

The adoption of the appropriate network strategy is important for opening and developing nascent markets. This strategy, even if successful should not be viewed immutable (i.e., think Kodak or Blackberry) and needs to be freshened-up as necessary to reflect the evolving needs of the ecosystem and the availability for improved tools for reducing market building friction.

The University Technology Supplier Network

There are approximately 15,000 universities in the world[7] and of these about 3,300 publish 80 % of the world's top peer reviewed research. Collectively, these institutions represent trillions of dollars in research infrastructure and ongoing research funding. Collectively, they produce more than 60,000 new discoveries per year in all areas of science, technology, and medicine. Together they constitute the leading idea factory of the world. Not to tap them for competitive technology intelligence, technology acquisition, and licensing opportunities is simply wasteful.

New technology development is a pluralistic sport, with components of a comprehensive solution often coming from several dispersed sources. This underscores the importance of creating and interrogating a network of discoveries to identify the relevant intellectual capital.

Social network theory describes both nodes and their ties. Nodes in this example are the technology transfer offices within the network and ties are the relationships between the nodes. In a more comprehensive view of discovery

---

[5] Gross (2003).

[6] UTEK Corporation.

[7] SIR Wold Report 2012, Global Ranking.

networks, the intellectual capital of the nodes can also be included and represented. In fact, the relationships between nodes of intellectual capital may be as or even more important than the strength of a specific node in generating go-to-market value from the network.[8] As Cook and Brown[9] have described, organizational knowledge is created through the interaction among sources of knowledge or discovery and those companies that have the valor to select sagaciously and can assemble a workable solution from existing discoveries. This interplay of knowledge and knowing is the raw material for new technology development in companies, based on exogenous discoveries and is defined as the structural intellectual capital of the firm.

The reason why the garage entrepreneur is less relevant today for the development of new discoveries is not due to a lack of brain power or funding but rather the heady advantage that social discovery networks can deliver to invigorate new product development. Knowledge is the most strategically significant resource for creating a competitive advantage for companies.[10] While ephemeral on one level, it may be extremely difficult to duplicate and in being so serve as a deep competitive moat.

As the world becomes more networked, the number of interactions between nodes grows exponentially. According to the Barabási–Albert (BA) algorithm,[11] the better connected a node the more likely it will be the beneficiary of new additions to the network (the so called preferential attachment, where P is the probability that a new node will be connected to node i as a function of the degree k_i of node i).

$$P \sim \frac{k_i}{\sum_i k_i}$$

Connected nodes are opportunities for sharing information and in a network of research institutions these data are discoveries. An offline, approach to technology development, however, well intended, is quixotic at best and will likely be wasteful of shareholder resources. Simply said, tapping networking research institutions worldwide, can provide a level of robustness for a corporate new technology pipeline that few, if any, in-house R&D efforts could match. To operate a business on the technology frontier requires working in part exogenously, with one foot squarely situated in the needs of your customer and the other in the networked world of innovations, relevant to your future business. If this balancing act is executed adroitly, both customers and shareholders will show their appreciation.

---

[8] http://www.fsc.yorku.ca/york/istheory/wiki/index.php/Social_network_theory.

[9] Cook and Brown (1999).

[10] http://www.fsc.yorku.ca/york/istheory/wiki/index.php/Knowledge-based_theory_of_the_firm.

[11] http://www.scholarpedia.org/article/Scale-free_networks#The_Barabasi-Albert_model.

References

Cook SDN, Brown JS. Bridging epistemologies: the generative dance between organizational knowledge and organizational knowing. Organ Sci. 1999;10(4):381–400.

Granovetter M. The strength of weak ties. Am J Sociol. 1973;78:1360–80.

Gross C U2B: a new model for technology transfer. BioEntrepreneur. 2003;21.

Peng M, Zhou J. How network strategies and institutional transitions evolve in Asia. Asia Pac J Manag. 2005;22:321–36.

Ventresca M. Handout #4 for strategy and innovation course, EMBA7. UK: Said Business School, Oxford University; 2012. p. 20.

Chapter 5
The Role of Design in Bringing Innovation to the Marketplace

Design is the sticky glue that holds and provides context for the multiple disciplines required to develop solutions to the difficult problem of addressing unmet, poorly articulated customer needs. Design may serve as a proxy for innovation or subsume it, depending on the vantage point selected. In both cases, it has the power to drive the firm forward and become the platform for its growth strategy. Not just design but 'design thinking' is able to address the anti-innovation, ingrained behavior expressed by some firms, as an antidote to inertial-laden, risk mitigation thinking, as an improved vehicle for driving corporate progress. In this sense, it is the new frontier for creating a competitive advantage, especially when capital, labor, and a continuous influx of external innovations are readily available. Economically, it resides within total factor productivity (TFP), the special sauce that improves the efficiency of companies and countries as they seek to compete on a global scale.

> Design Thinking refers to the methods and processes for investigating ill-defined problems, acquiring information, analyzing knowledge, and positing solutions in the design and planning fields. As a style of thinking, it is generally considered the ability to combine *empathy* for the context of a problem, *creativity* in the generation of insights and solutions, and *rationality* to analyze and fit solutions to the context.[1]

According to Kelley and Van Patter (2005) "as companies become more design sensitive their ability to innovate improves as does their economic performance." While anecdotal, and difficult to prove, it follows logically. Van Patter mentions that "Design, acting as the 'glue'—the bridge, facilitator, protector, explainer, valuer, modeler, orchestrator, and advocate of all thinking types."[2]

Not everyone agrees with the importance and utility of Design Thinking as a business tool. Nussbaum (2011) argues that "The decade of Design Thinking is ending and I, for one, am moving on to another conceptual framework: Creative Intelligence, or CQ." His thesis is based on the idea that Design Thinking is a process improvement and many companies have adopted it in a Kaizen-like manner of incremental, continuous improvement. As a result this has removed the

---

[1] Design Thinking (2012).
[2] Mentioned in Kelley and Van Patter (2005).

C. M. Gross, *Too Good To Fail*, Management for Professionals,
DOI: 10.1007/978-3-319-00281-1_5,
© Springer International Publishing Switzerland 2013

true look forward aspect of Design Thinking, namely to start at the solution and work backwards in a reverse iterative manner, but with a key benefit of starting quite a distance from the present situation. This is then followed by the sagacious selection of a winner from a number of alternative scenarios by present valuing the future user experience. In short, Nussbaum opines that the promise of Design Thinking has been beaten out of the paradigm by an overly structured approach to its implementation. He concedes that there have been some notable successes but many more failures. While Nussbaum might be right, beautiful products like the IPhone, IPad, Tesla Roadster create their own enthusiasm which extends their market penetration and shelf life.

Implications for Companies

More than 'glue' Design Thinking serves as an amalgam of many tributaries including, market research, financial modeling, science and engineering innovations, production, sourcing, human resources, and sales and marketing (Martin 2009). The big idea is to create a sustainable, competitive advantage through a design perspective that enhances the customer experience. According to Brown (2008), design thinking incorporates empathy, integrative thinking, experimentation, collaboration, and the optimism that a solution is waiting to be uncovered.[3] The idea of "form preceding function," is particularly relevant to the experimentation incorporated in Design Thinking. In a sense, this iterative design process is an analog of sped-up evolution. Like all evolutions, you begin with the solution and work your way up the food chain. In business the food chain is return on invested capital (ROIC). Make a better chopper or better yet make a better chopper experience and your ROIC will be greater than other companies in your neck of the jungle.

Commentary on Design's Role in the World

Design is fundamental to human experience. Whether you argue that form follows function[4] or fantasy, the design of forms that enhance the human experience is a large part of what separates man from other intelligent mammals. In fact, in nature, "form precedes function,"[5] with natural selection determining which forms are more durable. We are a species that has thrived in large measure due to our ability to make tools, e.g., consistent with the evolutionary selection of the opposable

---

[3] Brown (2008).

[4] http://en.wikipedia.org/wiki/Louis_Sullivan- (originally "form ever follows function").

[5] http://en.wikipedia.org/wiki/Form_follows_function

Fig. 5.1 Homo habilis (reconstruction from the West Fälisches Museum für Archäologie, Herne) (http://en.wikipedia.org/wiki/Homo_habilis) master stone tool user

thumb[6] or even five versus six fingers (polydactyl occurs in about one of every 500 births).[7] From the earliest choppers to MRI machines and supersonic aircraft, the channeling of human brain power and dexterity into and through tools has advanced the species (Gross 1996 and many others). Tools by definition provide an extension and amplification of human abilities, as such the interface between the tool and its user is often the limiting factor in determine a tool's effectiveness. The Oldowan[8] stone tool industry began approximately 2.6 million years ago in Ethiopia,[9] with stone and bone implements manufactured by Australopithecus and Homo Habilis (Fig. 5.1), and is the oldest record of human tool usage.

Although Homo Habilis was a regular on the menu of the large cat Dinofelis,[10] their sophisticated use of Oldowan tools is credited for their survival.

The history of product design could rightly be called the history of tool design. There is little doubt that the most influential tool was the book, the first tool that leveraged disparate knowledge and experiences tethered to a physical object. It served as the touchstone for all future tool developments by imbuing a designed object with time constrained information. More laconic and beautiful than any Bauhaus design, it transmuted ink and paper into a vehicle for sharing experiences, dreams (and nightmares), while simultaneously allowing for their storage for future reference. Of course these wonderful products have recently been liberated from their physical manifestations, gone are the paper, ink, and time limited information. In the current generation, books have become analogues for the continuous communication and sharing of human experiences. We have woven

---

[6] http://en.wikipedia.org/wiki/Thumb

[7] http://en.wikipedia.org/wiki/Polydactyly

[8] http://en.wikipedia.org/wiki/Oldowan

[9] Vrba et al. (1994)

[10] http://en.wikipedia.org/wiki/Dinofelis

	Design thinking as a cognitive style	Design thinking as a general theory of design	Design thinking as an organizational resource
Key texts	Cross 1982; Schön 1983; Rowe [1987] 1998; Lawson 1997; Cross 2006; Dorst 2006	Buchanan 1992	Dunne and Martin 2006; Bauer and Eagan 2008; Brown 2009; Martin 2009
Focus	Individual designers, especially experts	Design as a field or discipline	Businesses and other organizations in need of innovation
Design's purpose	Problem solving	Taming wicked problems	Innovation
Key concepts	Design ability as a form of intelligence; reflection-in-action, abductive thinking	Design has no special subject matter of its own	Visualization, prototyping, empathy, integrative thinking, abductive thinking
Nature of design problems	Design problems are ill-structured, problem and solution co-evolve	Design problems are wicked problems	Organizational problems are design problems
Sites of design expertise and activity	Traditional design disciplines	Four orders of design	Any context from healthcare to access to clean water (Brown and Wyatt 2010)

Source: Lucy Kimbell

Fig. 5.2 Three lenses from which to view design thinking (from Kimbell 2011)

books into the fabric of human consciousness; in doing so they have paved the way for all designed objects to travel the trajectory from things to experiences. All of these experiences require design to optimize usefulness and enjoyment.

With regard to business activities, an important subset of human activities, design plays the same role it has always played: making sure the designed experience (product and/or service + knowledge) is simultaneously transparent to the user, effective, and enjoyable.

Kimbell (2011) has summarized three lenses for viewing and studying Design Thinking (Fig. 5.2) [11] as "cognitive style," "a general theory of design," and "an organizational resource." Kimbell (2011) concludes that most claims as to the benefits of Design Thinking are anecdotal and require more systematic study to be better understood and made more useful. Specifically, she opines "the practices of designers play important roles in constituting the contemporary world, whether or not 'design thinking' is the right term for this".[12]

---

[11] http://www.designstudiesforum.org/journal-articles/rethinking-design-thinking-part-i-2/

[12] http://www.designstudiesforum.org/journal-articles/rethinking-design-thinking-part-i-2/

3M (Where Design is the Innovation)

Founded in 1902 and based in Maplewood, Minnesota, with more than $30B in sales and 84,000 employees; this remarkable company develops and markets more than 50,000 products.[13] Continuous innovation has been at the center of 3M's growth since the beginning. It is estimated that 1/3 of its sales comes from products that are less than 5-years old.[14] 3M has more than 7,000 full-time personnel dedicated to R&D[15] supporting over 40 research platforms. The company describes itself as follows:

"3M captures the spark of new ideas and transforms them into thousands of ingenious products. Our culture of creative collaboration inspires a never-ending stream of powerful technologies that make life better. 3M is the innovation company that never stops inventing."[16] Many of 3M's products have received international design awards. According to Mauro Porcini, the head of global design @3M "3M is committed to elevate design as a strategic innovation priority." 3M recently received 13 awards for design excellence including seven 2012 Red Dot Design Awards and six, 2011 Good Design awards[17] (Fig. 5.3). Design is literally and figuratively the "glue" that holds this Fortune 500 Company together.

Interestingly, the design and innovation leadership which has produced scotch tape®, post-it notes®, and scores of other award winning products was jeopardized with the implementation of rigorous six-sigma process improvement, during the CEO tenure of James McNerney[18] a protégé of former GE CEO, Jack Welch. What worked so well for GE at reducing waste and product variability cut to the quick of 3M's innovation and design machine.[19] Exceptional design and Design Thinking both require an ecosystem that tolerates ambiguity and failures (e.g., waste) as necessary iterations for achieving product greatness (Kimbell 2012). In a Mel Brooks-like manner, success is often built on multiple failures.

---

[13] http://en.wikipedia.org/wiki/3M

[14] http://www.businessweek.com/stories/2007-06-10/at-3m-a-struggle-between-efficiency-and-creativity

[15] http://www.microsoft.com/casestudies/Case_Study_Detail.aspx?CaseStudyID=4000005768

[16] http://solutions.3m.com/wps/portal/3M/en_WW/About3/3M/

[17] http://www.idsa.org/3m-wins-multiple-awards-design-excellence

[18] http://www.businessweek.com/stories/2007-06-10/at-3m-a-struggle-between-efficiency-and-creativity

[19] http://www.businessweek.com/stories/2007-06-10/at-3m-a-struggle-between-efficiency-and-creativity

Fig. 5.3 Filtrete water
station, winner of the 2011
Good Design Award and
2012 Red Dot Award (http://
www.idsa.org/3m-wins-
multiple-awards-design-
excellence)

Bang & Olufsen (Where Design Exists Primarily Without Technological Innovation)

Bang & Olufsen describes its heritage as follows: "We recognize the importance of identifying, cultivating, and exploiting our strengths. They are the glue that binds us together as a company as well as an engine of our business."[20] From humble beginnings in Struer, Denmark in 1925, Bang & Olufsen became the iconic brand for smartly designed radios, televisions, music systems, telephones, and loudspeakers. Their products are sold in more than 1,000 stores in over 200 countries.[21] Their self-described raison d'être is "Bang & Olufsen exists to move you with enduring magical experiences."[22] Their forward looking, sleek designs have positioned the company at the high-end of the consumer electronics market and captured many design awards including the 2012 Red Dot Award for its *Beolit* 12[23] (Fig. 5.4) portable music system which allows the user to play music from the IPhone, IPad, or PC, wirelessly.

For Bang & Olufsen, design is not an afterthought but the primary driver of its unique products. Since its earliest days it has leveraged external design expertise and manifested these designs in ultra-high quality products. This strategy has

---

[20] http://www.bang-olufsen.com/en/the-company

[21] http://www.bang-olufsen.com/en/the-company/heritage

[22] http://www.bang-olufsen.com/en/the-company/heritage/our-values

[23] http://www.bang-olufsen.com/the-company/press

Fig. 5.4 Bang & Olufsen's Beolit 12 portable music system (http://www.bang-olufsen.com/en/sound/sound-systems/beolit-12), winner of the 2012 Red Dot Award

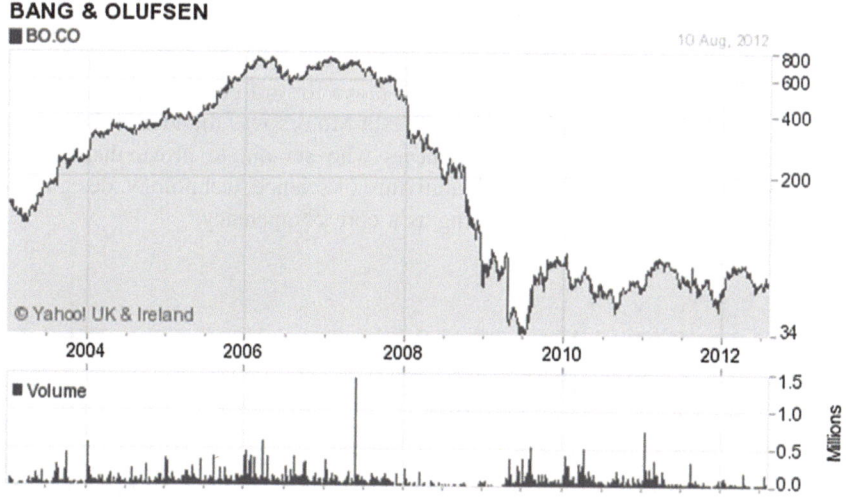

Fig. 5.5 Bang & Olufsen share price history (http://uk.finance.yahoo.com/q/bc?s=BO.CO&t=my&l=on&z=m&q=l&c=)

worked well for most of the Company's history. Beginning however in 2007, with the double threat of the growing popularity of consumer electronics products that leverage information sharing (e.g., IPod, iTunes, IPhone) as well as the economic recession which has understandably shunted customers away from premium priced products, has levied significant damage on the Company (Fig. 5.5). Since 2007 the Company has lost about 90 % of its market capitalization. While both individuals and companies have limited life spans, it hastens one to think that perhaps even a company with an outstanding ability to design products needs to pay equal attention to the design of business models and advanced technology to prevent rapid disintermediation. Without having developed improved business models, the beautiful designs of Bang & Olufsen have become less relevant for the creation of "magical experiences" for consumers and shareholders.

Tesla Motors (Where Design is the Package Around Innovation)

Telsa Motors is the first car company to produce an all-electric, lithium-ion powered sports car,[24] fusing the clean-tech movement with inspired design (Fig. 5.6). Founded in 2003 by Elon Musk and his team, they took the desire for a zero carbon footprint vehicle and wrapped in a forward looking, sexy design package. Conserving energy could now simultaneously be morally and esthetically beautiful and with 2X the energy efficiency as the popular Toyota Prius.[25]

Tesla has won numerous design and environmental awards for their beautifully designed electric vehicles and their charismatic and gifted CEO and Chairman Elon Musk received the 2010 Automotive Executive of the Year Innovator Award for inspiring the entire US auto industry to move toward electric vehicles.[26]

The company has a simple mission: "Tesla Motors was founded in 2003 by a group of intrepid Silicon Valley engineers who set out to prove that electric vehicles could be awesome".[27] With a mixture of science, technology, design, and style, Tesla embodies Design Thinking as a core competency.

Fig. 5.6 Tesla roadster (http://www.teslamotors.com/about) (photo courtesy of Tesla Motors)

---

[24] http://en.wikipedia.org/wiki/Tesla_Motors

[25] http://en.wikipedia.org/wiki/Tesla_Motors

[26] http://www.theoaklandpress.com/articles/2010/02/18/business/
doc4b7d1dd919bac072273954.txt

[27] http://www.teslamotors.com/about

Fig. 5.7 Dyson air-blade
hand dryer (http://
www.infolink.com.au/c/
Dyson-Appliances-Aust/The-
Dyson-Airblade-from-Dyson-
Appliances-The-Fastest-
Most-Hygienic-Hand-Dryers-
p16211)

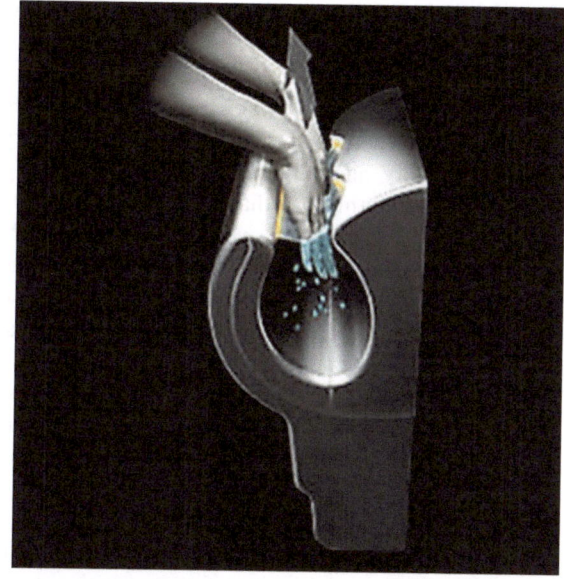

Dyson (Where Designing New Metaphors for Traditional Products is the Competitive Advantage)

Dyson was founded in the UK in 1992 by James Dyson[28]; the Company makes vacuum cleaners, hand dryers, fan, and heaters. While all of its product categories may be considered traditional if not downright stodgy, Dyson's take is anything but. The company is known for reinventing existing products to improve the user experience. It is reputed that Dyson's first product, a cyclone vacuum cleaner without a bag, was the end-result of 5,127 prototypes.[29] In 2009, Dyson won four Red Dot Awards[30] for the design of two of its vacuum cleaners and a unique airblade hand dryer (Fig. 5.7).

At Dyson, Design Thinking is the cornerstone of its new product strategy. James Dyson's training is unusual for an inventor or a CEO as he studied art and architecture at the Byam Shaw art school in London.[31] To reinforce Dyson's approach to Design Thinking, its founder has established the James Dyson Design Award.[32] According to the Company materials "The annual James Dyson Award

---

[28] http://en.wikipedia.org/wiki/Dyson_(company)

[29] http://en.wikipedia.org/wiki/Dyson_(company)

[30] http://www.infolink.com.au/c/Dyson-Appliances-Aust/Dyson-Wins-Four-2009-Red-Dot-Awards-for-Product-Design-n844678

[31] http://content.dyson.com/insideDyson/
article.asp?aID=jamesdyson&disType=&dir=&cp=&hf=&js=

[32] http://content.dyson.com/insideDyson/article.asp?aID=jda&hf=&js=

inspires and encourages students from around the world to engineer a solution to an everyday problem."

Perhaps the most well-known example of Design Thinking in action is that of Thomas Edison, whose invention of the light bulb through countless iterations, the power plant to generate the electricity and the delivery system requiring the wiring of an entire city and ultimately the nation; exemplifies the power of Design Thinking to transform more than just products and businesses but the entire human experience. It would be no overstatement, to say that the durability of General Electric (founded in 1892) which was added to the Dow Jones Index in 1896 and is the only company out of the original 12 that formed the index, to still be included as part of it today,[33] has resulted from Edison's Design Thinking. Of all of Edison's remarkable inventions, perhaps the one that has had the greatest impact is the modern corporate R&D laboratory, which he developed in Menlo Park. In 2012, GE was the third largest firm in the world[34] with revenues of $147.3B and a market value of $407.4B and a great example of creating marketplace value from design and innovation.

Design Thinking has been associated with corporate success, yet with few exceptions this success is fragile and highly dependent on the creative genius responsible for organizing, inspiring, producing and executing the products, systems, and experiences that incorporate this thinking. It is well known that significant brain power is not only precious and rare but exceedingly difficult to scale. While much can be learned by studying Design Thinking and design thinkers, it remains a humbling challenge to replicate these successes without gifted, visionary leadership. Having a great product innovation process is undoubtedly helpful, as there is a need to continuously fuse new technology to winning designs. Portions of this process may be acquired exogenously, for example, through technology sourcing as in the case of Tesla Motors, or design sourcing as in the example of Bang & Olufsen, but the guiding hand in these endeavors has to be inspired, chosen wisely, and be able to execute as though the future of the company depended upon it—because it does.

References

Gross C. The right fit. Oregon: Productivity Press; 1996.

Martin R. The design of business: why design thinking is the next competitive advantage. Boston: Harvard Business School Press; 2009.

Kelley D, Van Patter GK. Design as glue. Understanding the Stanford D. School. NextD Journal Conversation 21 http://issuu.com/nextd/docs/conv21); 2005.

Design Thinking, Wikipedia: http://en.wikipedia.org/wiki/Design_thinking, Accessed Aug 9 2012.

---

[33] http://en.wikipedia.org/wiki/General_Electric

[34] http://en.wikipedia.org/wiki/Forbes_Global_2000

Brown T. Design thinking, Harvard Business Review, June 2008. pp. 84–92.

Kimbell L. Rethinking design thinking. Part I Des Culture. 2011;3(3):285–306.

Kimbell L. Design thinking/design-as-practice, Handout, Session #3, MBA elective in designing better futures, Said Business School, Oxford University; 2012. p. 19.

Nussbaum B. Design thinking is a failed experiment. So what's next?, Co. Design, Editor Cliff Kuang, April 6 2011. http://www.fastcodesign.com/1663558/design-thinking-is-a-failed-experiment-so-whats-next, Accessed 11 Aug 2012.

Vrba J, De Heinzelin J, Schick K, Hart W, White T, Woldegabriel G, Walter R, Suwa G, et al. African Homo erectus: old radiometric ages and young oldowan assemblages in the middle Awash Valley, Ethiopia. Science. 1994;264(5167):1907–9.

Chapter 6
Leadership Required for Embracing New Technologies

New technologies are essential to demarcate a market and gain a competitive advantage, but their successful execution requires inspired leadership. The impact of the CEO on an organization's culture, perhaps more than any other feature, helps to determine the marketplace success of the firm.

The Impact of CEO's and Employees on Organizational Culture

Corporate culture, like all organizational attributes, imbues a corporation with both positive and negative inertia. In the best case, it can assist an organization in achieving growth, profitability, and efficiency, at worst, it can be debilitating and directly act in opposition to needed adaptation and change. Either way, as Pascale et al. (1977) noted, it is the "800 pound gorilla" in the room and it must be deconstructed and understood in advance of any serious "revitalization" effort.

Not Only the Good Die Young

Over a 40-year period from 1957–1997, nearly 85 % of the S&P 500 was replaced by newer entrants.[1] While some of these organizations have been acquired, many simply lost their competitive edge and either went out of business or have become significantly reduced in their size and relevance to their industries. One can view the effect of culture on an organization either prospectively or retrospectively through post mortem analysis. Organizations like Polaroid which seemingly did most things right for 40 years (or General Motors for nearly 100 years) can suddenly find themselves in bankruptcy and bordering on extinction. According to Nunes (2011) approximately 65 % of companies that hit a growth wall are either

---

[1] White (2011).

C. M. Gross, *Too Good To Fail*, Management for Professionals,
DOI: 10.1007/978-3-319-00281-1_6,
© Springer International Publishing Switzerland 2013

bought, go bankrupt, or are taken private. Their study assessed financial performance of companies over a 10-year period and revealed the somewhat obvious fact that the ones that performed well were able to reinvent themselves. Using observational techniques they identified three factors which they believe explain enhanced performance: anticipating the direction of the market, evolving competences, and commitment to a deep talent pool (human intellectual capital). While difficult to argue against the aforementioned virtues, on the surface it seems to be lacking a key element.

Leveraging Free Will: A Necessary Tool for Changing Organizational Culture

Large companies fail for the same reason they succeed—hubris. Henry Ford said, "If you think you can do a thing or think you can't do a thing, you're right."[2] If however, you think you can and you can't you are not only wrong you may damage the organization beyond repair. While this is somewhat simplistic, the will to succeed and the will to fail are inextricably interwoven into the culture of an organization. Two U.S. presidential elections ago, Barak Obama energized his fragmented base with the slogan "Yes we can." Three words that probably had the most enduring impact on determining the U.S. presidency since George Washington turned down the offer to be king of the new republic. The articulation and manifestation of the will of the people may, more than any other variable, affect an organization or even a country's destiny. The reason for this is that human will is the source of an individual's power to act. According to the Zohar (Berg 2002), free will is our most precious possession as well as perhaps our most powerful tool. Shaping, harnessing, and inspiring the will of an organization is equivalent to grabbing its cultural aorta and putting it to task. This begins with the CEO (the action) and reverberates across all of the employees of the organization (the reaction). When sagacious and purposeful, the channeling of the CEO's will can inject the activation energy into an organization, by inspiring all of its members to act in a coherent and positive manner; and in doing so overcome the hubristic inertia that is preventing it from adapting to the current conditions.

Corporate inertia is not always a bad thing and may even serve well for constant velocity sojourns. However, when abrupt changes in direction are required, an entrenched corporate culture can result in high "g" forces rendering employees non-compliant or marginalized. Pascale et al. (1977) thesis is that employees at all levels of an organization need to be passionate and proactive for corporate improvement programs to produce meaningful change. Pascale refers to this type of movement as "revitalization" (Pascale et al. 1977) and shows to good effect how it was achieved at Sears, Shell, and the U.S. Army by their respective CEOs;

---

[2] The Quotations Page (2011).

directly reshaping the culture of their organizations. Pascale et al. (1977) have suggested that a quantum leap is required in corporate performance to make significant change and that there are three agents required to achieve this change "incorporating employees," "leading from a different place", and "instilling mental disciplines." Pascale et al. (1977) further identified four variables that they believe are predictive of success "power," "identity," "conflict", and "learning." The upshot being that meaningful change, at least in the case of Sears, Shell, and the U.S. Army occur from the bottom-up. The challenge is getting top management to deliver a sufficient dose of activation energy to the troops that is able to overcome cultural inertia and get the wheels rolling in the right direction. This is underscored by Pascale et al. (1977) observation from Nitin Nohria's work (1977) that approximately 70 % of the cultural change programs implemented in Fortune 100 companies over a 15-year period, did not produce a bottom line result sufficient to cover the cost of the improvement programs. This is both unfortunate and telling as Pascale et al. (1977) reported that the average program costs roughly $1 billion (excluding opportunity costs which might be significantly higher). Pascale et al. (1977) hypothesized that the high failure rate of cultural change programs results from "...the whole burden of change typically rests on so few people." While Pascale makes a good case, it is far from certain that the will of the group or the *tyranny of the masses* (De Tocqueville 1838) is the panacea. Few would argue against the notion that management's toleration of ambiguity is a strength, however, the bottom-up ability to change a corporate culture may also lead to a new category of problems when disruptive innovation is needed, e.g., Sony versus Apple in the rationalization of how music is purchased and played. The harnessing of friendships based on a common enemy may get the job at hand done, but may turn counterproductive once either the enemy is subdued or enthusiasm for the campaign has waned.

Social Engineering Applied to Corporate Culture

Goffee and Jones (1996) observe that "Culture, in a word, is community. It is an outcome of how people relate to one another." Goffee and Jones (1996) further suggest that you could measure the integrity or strength of the community using the cultural scalars "solidarity" and "sociability." Sociability is commonality based on friendship, while solidarity is commonality based on purpose. It is important to note that solidarity is a priori neutral. Only when imbued with a positive ethos, does solidarity take on a positive connotation. With this nomenclature Goffee and Jones (1996) describe simplistically, four theoretical types of communities, "networked", "mercenary", "fragmented", and "communal." In fact, it is highly likely that any given organization is an amalgam of all four cultural types to a greater or lesser degree. Irrespective of the criticism, Goffee and Jones (1996) reasonably assert that an organization should seek to have a type of culture suitable for their environment. When there is a mismatch, such as when a communal culture needs a sharper edge to regain a lost competitive advantage or even for survival, then one could visualize the need to shift the culture to the mercenary type, for example. The compelling description of Timmer's (CEO of Phillips Electronics) efforts to transmute Phillips from a networked to a mercenary culture demonstrate an enthusiasm for cultural engineering but is itself not generalizable. This example does however underscore the ability of CEO's to orchestrate and implement cultural change programs. How long such changes last, how effective they are in the short-term, and whether they generate a return on investment would need to be quantified to determine if it is worth the carriage. The interesting take away is that management can modify an organization's culture using the levers of sociability and solidarity, pushing it from one quadrant to another to better position it to meet the current needs of the organization, in the spectrum between self-actualization of employees and optimized profitability of corporations. Many have chided behavioral science as a soft science but this may be a new anchor tenant. Goffee's (1996) paper is purely anecdotal, yet with earnest and interesting observations. Yes, corporate culture can be changed by deterministic actions by CEOs, but to what extent, are the changes durable and is it worth the cost?

By contrast, Schein (1996) believes that organizations "were much more likely to change leaders than be changed by them." A priori, logic would dictate that if managers could change organizations, organizations could equally well change managers depending upon specific facts and circumstances, i.e., relative inertia versus required management expertise and activation energy. In this manner, Shein shares Meek's (1988) view as described by (Dopson 2011) that culture is not an aspect of a company but rather the company itself. As a result of his introspection, Schein (1996) posits that perhaps a company is not the best unit to study in organizational psychology, not dissimilar from trying to look at a large painting up-close. You see everything and nothing, as if the focal point is behind you. Schein (1996) provides a pretty compelling rationale for this problem by stating that corporate culture is not one entity but rather an amalgam of three sub-cultures consisting of "operators," "engineers," and "executives." These three sub-cultures are interwoven into a corporate constellation without a defined or easily determined centroid. This has powerful implications for assessing the value and relevancy of hierarchical systems such as Goffee's (1996) communities and Pascale et al. (1977) change agents for analyzing corporate cultures, i.e., caution as to using smart, statistically measurable abstract quantities that do not have face validity. Schein (1996) suggests that we need to learn more about the cultures swirling around inside companies before we sell recipes for fixing them— hard to argue. He also strongly suggests that if we want to understand a corporate culture, we need to look first and suppress the reflex to measure, i.e., substance over form.

Can Leadership Inspired by and Receptive to Technological Innovation Reshape Organizational Culture?

My experience as CEO of two companies (both of which were publicly traded) over a 20-year period resonates with Schein's (1996) observations. In one of those companies, we provided product improvement services for many of the Fortune 500. I will discuss a few experiences we had with Black and Decker, Lear, Reebok, and Knoll to illustrate the benefit of Schein's advice, i.e., if you want to understand an organization's culture, look before you measure.

Reebok under the leadership of Paul Fireman had experienced unprecedented growth as baby boomers embraced fitness as a core value. Reebok needed a technological edge to be more competitive with Nike and Fireman was able to deliver. Reebok introduced the pump sneaker, thinking perhaps it would enhance the technical wow factor of the brand, by offering the first running shoe that was volumetrically adjustable with air. Initially, they planned for a 50,000 unit production run and launched the campaign. Within weeks following the launch, the roll out was so successful they had to increase the first production to 500,000 units. The pump shoe was everywhere and the media and sports aficionados alike were thrilled that a new adjustable dimension was finally added to the somewhat stodgy athletic shoe. But where to go next? We approached Reebok with an upgrade strategy, the smart pump; a microprocessor controlled pump shoe that automatically sensed the pressure on the foot and inflated or deflated corresponding bladders in the shoe using a gravity-activated heel pump. The R&D team at Reebok was excited and impressed and at their urging I met with Mr. Fireman to present the idea as well as the license agreement for the underlying technology we had invented to make it possible. Paul Fireman was not a technology expert but he had an uncanny sense about marketing products and the ability to think strategically and imaginatively about fostering and fulfilling consumer demand. Interestingly, the R&D team at corporate headquarters displayed a very similar penchant for embracing new technology to move their products forward in the ultra-competitive athletic footwear industry. Fireman had imbued the company with the edgy attitude that taking chances was the right thing to do to survive and prosper in the space. Reebok subsequently licensed the technology from us and when Adidas acquired Reebok in 2006, Adidas introduced the world's first smart shoe. Approaching problems with a new lens was viewed as the right thing to do by Reebok's line and management as a result of Fireman's urging.

The cordless power tool industry is a crowded space with many capable competitors with barely distinguishable products. In the 1990s, the Japanese cordless tool manufacturers like Panasonic had a competitive advantage over their European and American rivals in as much as they also supplied the industry with the rechargeable batteries, a fortuitous by-product of their experience in the consumer electronics space. I met with the charismatic, can-do, President of Black and Decker Tools USA and presented a strategy for Black and Decker to differentiate themselves, with a cordless drill that was less stressful and easier to hold than the competition. He seized upon the concept and immediately empowered and inspired an extremely capable new product team to make it happen. Using biomechanical models of the hand and wrist and unique instrumentation to measure muscle forces, we designed a new type of drill that minimized the torque on the wrist making it more comfortable to hold throughout the day. Church groups across the country were mobilized to blind test the new product against the leading selling tools from the competition. The results were extraordinarily positive for the new drill. Within a few months, the new DeWalt tool was in production (an ultra-short roll out) and subsequently became the leading selling cordless tool in the company's history and a favorite among journeymen worldwide. The product was awarded a gold medal from Business Week as the best product in its category in the following year and was placed in a Smithsonian exhibition on ergonomic product breakthroughs. As impressive as the product itself was the speed with which a somewhat traditional organization was mobilized into action around this new technological innovation, through the vision and drive of the division president. Using the catalyst of technological innovation, Black and Decker was able to leap frog its rivals with a superior product that was the fruit of a revitalized corporate culture put under harness.

Lear Seigler Seating Corp. (now known as Lear Corporation) was spun out of Lear Seigler Holdings Corp. in 1988 in what was up until that time the largest LBO in the automotive sector. Ken Way was the then CEO with a large task at hand; revitalize a dyed in the wool corporate culture while expanding the business for its new owners (private equity firm Forstmann Little and Company). Ken was a visionary who set out to reenergize Lear's seating business in its first incarnation as a standalone company. He built a new and impressive R&D center in Southfield Michigan and set its sights on winning more of the just-in-time seat supplier business from the major automobile manufacturing companies. Ken immediately grasped the benefit of self-adjusting automobile seat contours as a way to get a leg up on the competition offering simple lumbar supports for driver seats. He also embraced the notion of quantitative measurement of seat comfort as part of the design process. Ken incorporated the new technology for the automatic adjustment of car seats to optimize user comfort. A high end prototype was built and presented to Cadillac for comfort assessment. For nearly 100 years, Cadillac rated all seats for their vehicles based on extensive focus group trials. Seat were measured on a comfort scale of 1–5 with 5 being the most comfortable possible. No seat ever received a 5 in the history of Cadillac. The new prototype self-adjusting seat was tested and rated 5. Ken as CEO inspired product development to embrace the new technology to exceed customer expectations. The seat was introduced into the Cadillac brand a few years later and sales growth followed.

Knoll is the indisputable design leader in office furniture. Known as a design house first, Knoll has always prided itself with leading edge, enduring designs (think, Ludwig Mies van der Rohe and you get the idea). Office chairs more modestly priced were never their forte. Times changed and Knoll realized in the 1990s that they needed a price competitive chair with extraordinary ergonomic features to grow sales. We were engaged to develop a range of seat contours based on a scientific study using a new technology to measure pressure between the seat and the user. Knoll incorporated these contours into the Parachute chair, which following its introduction became the leading selling mid-priced office chair in the U.S. Having science-driven design was a paradigm shift for Knoll, only possible due to inspired management and staff that understood that their design culture had to change to remain competitive. Here was an example of the product development staff accurately assessing the limitations of the existing design culture and developing a work-around for the sake of corporate success.

The take-away from the above cases, while anecdotal, underscores how even tradition laden companies can reshape their cultures when the market demands it and if management is receptive, forward-looking, and inspirational. Underscoring this point is the change in corporate culture that was described by Morrison (2011) when referring to the change that had taken place in Goldman Sachs' corporate culture and perhaps the entire large capitalization investment banking community, from one of reputation-based client engagements to trading-based client engagements. Morrison (2011) further points out that the catalyst for this may have been a combination of industry trends due to technology improvements and regulatory oversight combined with the personality and predilection of Goldman Sachs' CEO, as a former securities trader. Here, the issue of a leader's impact on organizational culture is more complex, as one could argue that the culture writ large, namely the investment banking community had produced the change in the individual business due to the progress and availability of computer technology for instantaneously determined changes in the values of securities and options. Equally, one could also argue that in spite of this environmental change a specific organization is still under the guidance of its CEO and board in determining quantum cultural changes it may embrace as it executes its business model. According to Morrison (2011) Lloyd Blankfein clearly argued in his testimony before the U.S. Congress that Goldman Sachs was acting as a market maker, and therefore, acting properly within the confines of a trading type business with regard to the ABACUS transaction. The problem in this interpretation is the lingering artifact of the Goldman Sachs brand, which for the better part of a century has stood for reputational excellence above and beyond any specific transaction. In fact, the culture had changed, but who or what precipitated the change, the CEO, the industry, the regulators, the change from a partnership to a corporation, or an amalgam of these factors are all possible causative agents and exceedingly difficult to deconstruct and value independently.

Interestingly, in my judgment, the rally around new technology never gets old and has the ability to become a tool to facilitate organizational cultural alignment to market needs. Under the right circumstances senior management can effect cultural changes, but it does not happen often enough, or when it does, necessarily with good effect. Sometimes the inertia is just too great and companies continue doing what they previously have done well. In doing so, often they march with measured efficiency into the abyss of obsolescence, or worse, irrelevance. Sometimes organizations are up to the task of changing their cultures, but the type of change they implement may be counterproductive to their long-term goal of being a durable, profitable, well respected enterprise. CEOs may have a significant impact on a corporation's culture but as Barney (1986) intuited, it will only create sustainable superior financial performance if the culture being tweaked is VRI or as Barney (1986) has said, "valuable, "rare" and "imperfectly imitable." If true and extrapolative, Barney's work has interesting implications, in as much as leaders of firms without VRI cultural factors are likely not to gain much from trying to foster changes in the corporate culture and may be relegated to mediocre or worse positions absent wholesale changes that would render the organization VRI. While this is a reasonable proposition resulting from Barney (1986), it is by no means proven to be correct. Barney (1986) did not conduct a controlled study but rather reviewed the literature through the convex lens that focuses uniqueness at the focal point of superior financial performance. Additionally, even if it were true, it is unlikely in my experience to be true under all conditions and time frames. For example, markets efficiently price securities; however they do not do it instantaneously. The reason being is that the value of a security is an amalgam of its present value and its future potential. While the present value may be calculated and benchmarked the future value is subject to the vicissitudes of the market, the lens of the observer, the strength of the global economy, and a host of internal and exogenous factors that together are impossible to predict without a high degree of variance. This is in fact what makes investing interesting and risky; otherwise equities would be deconstructed into bonds. Ultimately, value is known, but between now and ultimately, a lot can and does happen. The same may be said of VRI cultural factors, ultimately they may be predictive of success but between now and ultimately, companies not possessing VRI cultural factors may succeed and those possessing them may fail due to more salient current issues that temporarily override theoretical considerations along with a host of other logical imperatives. However, one VRI factor that can be relied upon more than most is proprietary intellectual property, especially when sourced cost-effectively.

Theoretical Models for Organizational Culture: More is Less

To determine the extent to which both leaders and associates impact organizational culture, it would be useful to have a theoretical framework with which to view organizational culture. Hatch (1993) makes a good case for expanding Schein's (1985) model on cultural dynamics to include symbols as these are often touchstones of cultural foci. More important and modest is the suggestion that the linkage between the elements may contain more value than the elements themselves. Yet the model still appears flat, lacking face validity. Though well-articulated, Hatch's (1993) refined model is an abstraction of an abstraction. Organizational culture is a theoretic construct as are the imputed workings and interrelationships of Hatch's (1993) "values," "artifacts," "assumptions," and "symbols;" unsteady ground in my reckoning from which to draw a horizon line. Using Hatch's (1993) model to describe an organizational culture is no more accurate than looking at the inner workings of a Swiss watch to predict time. While it is certainly in the neighborhood, it is neither necessary nor sufficient for an accurate extrapolation of time. More is needed to create an objective framework that is anchored in quantifiable reality. To this end I propose a cultural dynamics model that incorporates Hatch's (1993) good work but expands it to include what I call Observable Fruits (Fig. 6.1). These fruits are of three varieties, financial performance, degree of market manifested innovation, and proven adaptability. By starting with the quantification of the Observable Fruits, it may be possible to anchor the model using normalized objective data which would thereby reduce the predictive variance of the underlying mechanism being used to produce the fruits. This would potentially solve a lot of problems, not least among them the mis- guided notion that according to Hatch (1993) in agreeing with Schein (1985) "that artifacts are the most tangible aspects of culture."

As artifacts are rearward looking, one could argue that they would take a back seat to financial performance as being a tangible measure of the performance of a culture. Using an artifact to measure cultural vitality is like using a snake skin to measure the health of a snake. It may provide some useful information, but a live body assessment would inherently be more valuable, albeit perhaps more difficult. This is however the method used by physicists to describe the forces that nature obeys, such as when planetary motion was first observed by Kepler using Danish astronomer Tycho Bryhe's data which described the orbit of Mercury. This was done before there was a good understanding of the forces necessary to produce

Fig. 6.1 Model for understanding the effect of cultural factors on observable corporate performance (Modified after Hatch 1993)

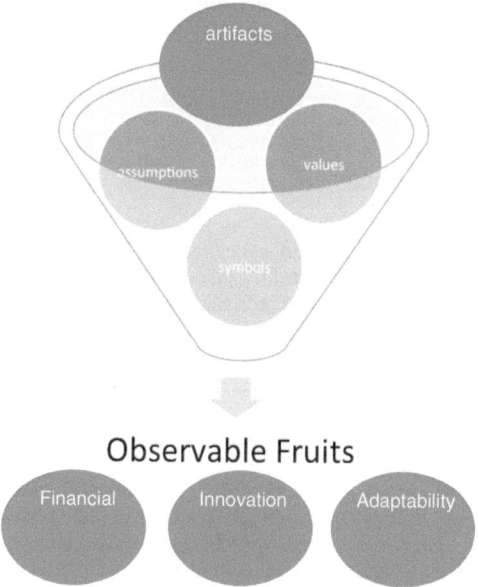

such motion. Understanding the mechanisms that produced the orbits would have to wait for Newton to theorize the law of universal gravitation. With a bit of grace, accurate observations can meet up with theory, although not necessarily in the same study.

The extent to which leaders and staff can create, modify, or follow organizational culture is highly dependent upon the facts and circumstances of the organization, its history, current viability, business climate, and caliber of staff and management. When there is a positive alignment of the market forces and people, a reshaping of the culture can occur, certainly for short periods of time. More probable perhaps, for a well-established, large organization, it would be relatively unyielding to the forces of management and even inspired employees due to the lack of sufficient activation energy, imparted by senior management, staff, and/or exogenous factors that are sufficient to produce needed cultural change. In such a case, even the effect of inspired management may be muted.

Technological innovation, especially disruptive innovation, has been observed as a potentially useful catalyst to inspire cultural change whether from the top-down or the bottom-up. This may occur by creating a level of enthusiasm which results from the visualization and game changing potential of an improved or superior product. The simple fact remains that it is more important what a company does than how it does it. Current social networks have taught us that no company exists in the confines of it corporate walls but is organized within the ecosystem of technology and the market. Rather than seeking to correct corporate shortcomings with changes in organizational culture a more direct solution would seem to be neutralizing the specific weakness with an exogenous strength. In the the arena of innovation this implies leveraging global networks of innovators to help ensure that the firm has the best possible products and services. It would be tough to argue that anything could be more valuable, rare or imperfectly imitable than proprietary (patented) technological innovation; especially if such innovations empower customers to be more productive.

References

Barney JB. Organizational culture: can it be a source of sustained competitive advantage? Acad Manag Rev. 1986;11(3):656–65.

Berg RPS. The essential Zohar. New York: Bell Tower; 2002.

De Tocqueville A. Democracy in America [Henry R, Trans.]. New York: George Dearborn and Company; 1838.

Dopson S. "Organisational culture" lecture, MBA program, Jan 2011, Said Business School, Oxford, UK.

Goffee R, Jones G. What Holds the Modern Company Together? Harvard Bus Rev Nov–Dec. 1996;134.

Hatch MJ. The Dynamics of Organizational Culture. Acad Manage Rev. 1993; 18(4):657–693.

Ford H. If you think you can do a thing or think you can't do a thing, you're right (The Quotations Page). 2001. http://www.quotationspage.com/quote/2330.html. Accessed 21 Jan 2011.

Meek V. Organizational culture: origins and weaknesses. J Organ Stud. 1988; 9(4):453–473

Morrison AD, Willhelm WJ. Computerization and the ABACUS: reputation, trust, and fiduciary responsibility in investment banking. Paper provided as part of the class notes for executive MBA program, Jan 2011, Said Business School, Oxford, UK; 2011.

Nohria N. From the M-form to the N-form: taking stock of changes in large industrial companies"(Harvard Business School Working Paper 96-054). As referenced in Pascale R, Millemann M, Gioja L. Changing the way we change. Harvard Bus Rev Nov–Dec; 1977.

Nunes P, Breene T. Reinvent your business before it's too late. Harvard Bus Rev Jan–Feb. 2011;80–87.

Pascale R, Millemann M, Gioja L. Changing the way we change. Harvard Bus Rev Nov–Dec. 1977;127–129.

Schein EH. Culture: the missing concept in organizational studies. Adm Sci Q 40th Anniversary Issue. 1996;41(2):231, 236.

Schein EH. Organizational culture and leadership. San Francisco, Jossey-Bass, 1985.

White A. Discontinuous innovation. Lecture in executive MBA program, Jan 2011. Said Business School, Oxford, UK; 2011.

Chapter 7
The Growth of China's Technology Transfer Industry

While peering into the future is not my forté, a description of the business of commercializing university discoveries to accelerate the growth of firms would be less complete, without a discussion of role China will likely play in this emerging industry.

China has moved mightily over the last 30 years to increase its capacity to develop indigenous technology to invigorate its industrial base and shift it from the world's factory to the world's developer and manufacturer of products. To achieve this requires buttressing an emerging intellectual property (IP) system, increasing university research while encouraging scientists to patent and commercialize their discoveries. Additionally, the development of a functioning and agile venture capital (VC) system to invest in these new technologies coupled with liquid equity markets for consummating IPOs have been developed in record time.

Will these remarkable efforts be sufficient to allow China to dominate the technology transfer market domestically or internationally over the next 10 years? The conclusions are twofold: on the one hand, China undoubtedly will become the world's largest customer for technology transfers both domestically and internationally, but global leadership in new technology development and licensing from research institutions is unlikely to be achieved over the next 10 years. Foreign firms, especially those within the US or with strong ties to the US, are most likely to dominate this sector due to the US's comprehensive university network coupled with its well-established IP technology transfer industry.

The major implication for the global technology transfer market is a combination of significantly increased demand which will be met with increasing numbers of transactions from both domestic and international suppliers of new university discoveries. As a result of China's rapid growth, the US and Chinese technology transfer market should be at least twice its current size in 10 years' time. This portends to a significant increase in the number of new university

Excerpted from: C. Gross, "The growth of China's technology transfer industry over the next decade: implications for global markets," Journal of Technology Transfer, Springer Netherlands, Doi: 10.1007/s10961-012-9263-x, August 2012.

C. M. Gross, *Too Good To Fail*, Management for Professionals, DOI: 10.1007/978-3-319-00281-1_7,
© Springer International Publishing Switzerland 2013

discoveries that will be brought to market to create economic value for both countries, while increasing the quality of life, productivity and the standard of living through the application of science and technology.

Technology transfer is concerned with bringing technologies from the source of innovation to the marketplace. The most vibrant source, worldwide of new innovations, is research universities. These idea factories conduct basic research in every area of science, medicine, and technology. When new discoveries are made, usually they are disclosed to the institutions technology transfer office (TTO), whereupon the technology is assessed with regard to its market and IP potential. In the case of a potentially breakthrough technology, intellectual property rights (IPRs) are applied for through the filing of patents.

Though generally recognized as essential for future economic growth, the technology transfer industry is immature, with the major actors being the research universities themselves. A few commercial enterprises have been established outside the university systems to facilitate transactions. The US is currently the world leader in technology transfer. This is the result of the forward looking Bayh-Dole Act of 1980, which for the first time, empowered research universities and Federal Research Laboratories to have ownership of their discoveries produced with taxpayer-funded research; empowering them to effectuate transfers, usually through licensing agreements to commercial enterprises. There have been some pretty dramatic successes as a result of university technology transfer including, Google, Genzyme, Gatorade and host of successful, and game changing products.

In 2000, China replicated the Bayh-Dole Act domestically with the hopes of developing an indigenous new technology pipeline to invigorate domestic industries. As a result, TTOs have been set up at major universities across the country to provide a network to facilitate and encourage transactions.

For China, this is no side bar effort. The nation's leadership is focused on improving the innovation capability of Chinese companies as early gains in gross domestic product (GDP) as a result of deploying capital, natural resources, and labor are unlikely to be replicated without significant future increases in Total Factor Productivity (U.S. Congress 1987), specifically intellectual capital (Wu 2010). It is well recognized (Orcutt and Shen 2010) that sustainable GDP growth in China is dependent of the innovation capacity of Chinese industry. There are three possible approaches to address this need for innovation:

- Indigenously development of innovations through expand basic research capacity.
- Exogenously acquired innovations through technology transfers from other countries that possess enhanced technology innovation capabilities.
- An amalgam of 2 and 3 above.

The economic battle ground for the next decade will be fought to capture and nurture the soil of innovation. The war that is coming is the war of big ideas, disruptive scientific discoveries. Growing these discoveries and harnessing them to accelerate the growth of industry is essential for economic prosperity in the coming decade for both China and the US.

It is likely that the winners of this battle will be those countries that pursue an amalgam of indigenous innovation and open innovation. The US has the most robust and largest intellectual capital infrastructure in the world with 2,000 colleges and universities and 700 federally funded research laboratories. All of whom have in varying degrees, the capability to generate novel discoveries as a result of pushing the research frontiers of science, medicine, and technology.

China is moving mightily as well. She has also established and funded a network of 2,000 universities, colleges and technical institutes, coupled with 30 science parks and more than 200 government operated research laboratories. To align incentives and replicate some of the successes of the US system, China has adopted Bayh-Dole like laws (Feng 2009) empowering these research institutions to own the IP developed from their research and commercialize their discoveries.

Having the means of production of innovation is necessary but not sufficient to guarantee economic leadership. The larger ecosystem for systematically developing innovative, world-leading industry requires a robust intellectual capital framework including a functioning patent law system for asserting and protecting IP rights. Additionally, the VC and investment banking industries have to function to encourage investment throughout the society not just government directed research funding.

From the *tableau raza* of 1980, China has established a patent office, and in 1984 enacted patent laws to respect and encourage the development, protection, and ownership of IP. The results in the subsequent period have been remarkable. China now has the world's most active patent office and in 2011 filed 26.4 % of the world's patents, eclipsing the US's 23.5 % contribution for the first time (Fig. 7.1).

IP filings by office and income group

Office and Income Group	Share in world total (%)						Average annual growth (%)		
	2008	2011	2008	2011	2008	2011	2008-2011		
	Patents		Marks (class count)		Designs (design count)		Patents	Marks	Designs
China	15.1	24.6	12.8	22.8	43.6	53.1	22.0	26.6	18.6
European Patent Office	7.6	6.7	n.a.	n.a.	n.a.	n.a.	-0.8	n.a.	n.a.
Japan	20.4	16.0	3.7	3.0	4.7	3.1	-4.3	-2.1	-2.8
OHIM	n.a.	n.a.	4.6	4.9	11.3	8.9	n.a.	6.7	2.4
Republic of Korea	8.9	8.4	3.7	2.8	8.2	6.0	1.6	-4.8	-0.2
United States of America	23.8	23.5	7.3	6.6	3.9	3.1	3.3	0.9	3.1
World	100.0	100.0	100.0	100.0	100.0	100.0	3.8	4.3	11.0
High-income	74.8	67.0	52.8	45.1	44.9	37.2	-0.3	-1.0	4.2
Upper middle-income	22.2	29.8	35.5	43.9	52.0	59.5	14.2	12.1	16.0
Lower middle-income	3.0	3.2	10.4	9.9	2.8	3.1	5.2	2.7	15.9
Low-income	0.1	0.0	1.3	1.0	0.3	0.2	-38.5	-2.4	-7.4
World	100.0	100.0	100.0	100.0	100.0	100.0	3.8	4.3	11.0

Note: OHIM = Office for Harmonization in the Internal Market; Trademark data refer to class counts, i.e., the number of classes specified in applications. Industrial design data refer to design counts, i.e., the number of designs contained in applications; n.a. = not applicable

Source: WIPO Statistics Database, October 2012

Fig. 7.1 IP filings by country and income group (WIPO 2012) (http://www.wipo.int/export/sites/www/ipstats/en/wipi/pdf/941_2012_highlights.pdf) *Source*: WIPO, World Intellectual Property Indicators 2012

Additionally, China has launched a VC industry, essential for the financing of new university discoveries and the businesses that seek to bring them to market.

To encourage venture investing, China has also re-established several stock exchanges (Ordish and Adcock 2008) and has vigorously encouraged the consummation of IPO's to expand its growing industrial base as well as to meet the appetite of a growing group of investors for this asset class. The Shenzhen Stock exchange was opened for trading on December 1, 1990 and the Shanghai Stock Exchange was re-established on December 19, 1990.

To better understand the ecosystem for technology transfer one needs to take into account the technology development efforts along with the IP landscape, the market for recipients of technology transfer, namely, corporations and the organizational factors which, for this industry, encapsulate technology transfer laws in the respective countries, both for indigenous growth and with regard to importing new technology.

Ventresca (2012) describes the system building activities necessary to bring new technologies to the marketplace (Fig. 7.2). Technology transfer plays a key role in enabling new discoveries to cross over from the land of big ideas (universities) to the world market.

From a national perspective, the ecosystem for commercializing discoveries cuts through the major societal institutions beginning with knowledge creation at educational institutions, patenting to stake a claim on the intellectual capital, technology transfer to bring the IP to a company, VC to finance the company and an effective public company or late venture marketplace to provide further financing and in the process monetize the VC investments (Fig. 7.3). This ecosystem requires a more or less level playing field of IP ownership, protection, and transactions to be effective.

System phase	Primary activities	Key actors	Supporting elements
Invention	Create raw 'novelty'; external to current common sense	Independent inventors unconstrained by organizations	Academics, vendors, corporate R&D, patent offices
Tech Transfer	Adapt and adopt technology: Impact on markets & industries	Context-specific actors 'customize' innovation in local value regimes	Political and legal environment; industry leaders
Development	Embed invention in business functions; recognize PESTEL	Inventor-entrepreneur or organization	Other professionals (e.g. engineers, scientists) and institution actors
Innovation	Produce value from emerging tech – value creation system (VCS)	Manager-entrepreneur (e.g., Schumpeter on entrepreneurs)	Incumbent and new entrants, new dominant designs
Consolidation, growth, and competition	Build system momentum; notice & resolve reverse salients	Finance-and policy-entrepreneurs; incremental innovation	System momentum makes tech appear inevitable trajectory

Fig. 7.2 System building activities (Modified from Ventresca 2012)

Fig. 7.3 The value chain for technology transfer

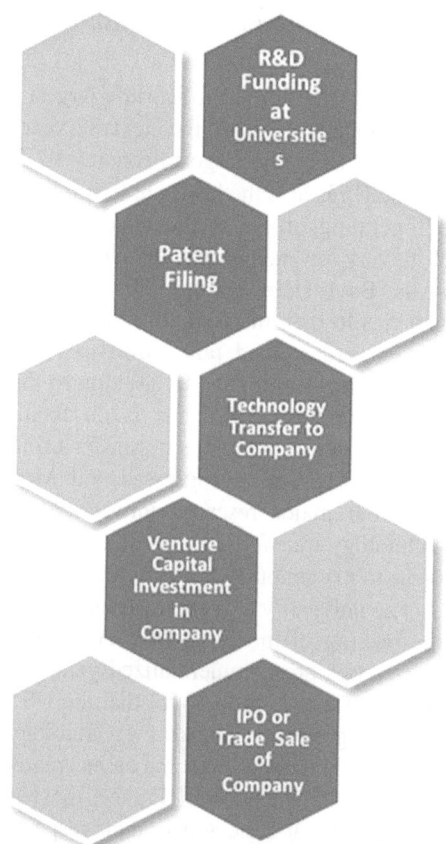

Since Henry Chesbrough's (2003) seminal work on open innovation, most technology transfer actors have become aware of the importance of the ubiquitous and unpredictable nature of scientific discoveries; usually the product of a few fertile minds, unbounded by geographic or corporate boundaries. Hence, the sourcing of discoveries for both companies and countries has become a truly global endeavor (Palfrey 2012). Although counter intuitive to some, "not invented here" is actually a virtue as it defenestrates the illusion of control and respects the fundamental truth that wherever you are, and whatever company you may be in, most new technologies are not invented there.

As the US has developed the technology transfer industry proper in 1980 and continues to dominate the industry, a comparison of the US and Chinese capabilities in this area are reviewed. This chapter is focused at understanding the dynamics of this growth and projecting which country is likely to assume the dominant role in this industry over the next decade.

The U.S. University Technology Transfer Market

Currently, the US is the world's largest economy by GDP, although this is projected to change over the next 38 years (Morrison 2011) whereupon China is projected to be the largest economy by 2050 if not sooner.

The US is also the leader in technology transfer. The industry consists primarily of operating offices at the major universities and federal research laboratories. With very few exceptions, these offices were established as part of the requirement of the Bayh-Dole Act in 1980. This unique piece of legislation empowered universities to own the fruits of federally funded research with the proviso that they put forward a good faith effort to commercialize these discoveries; hence, the establishment of TTOs. In addition to these offices, there are a handful of private consultancies whose focus is on facilitating technology transfer. Financially, however, these private tech transfer firms are small relative to the overall market, although a few have progressed well. Most notable technology transfers have come from the major research universities. According to Gross (2003), university technology transfer is a potential engine room for small cap growth by enabling them to exogenously acquire new discoveries with an open innovation platform.

The university technology transfer market has expanded greatly in the US since the passing of the Bayh-Dole Act. Although only 22 % of disclosed university discoveries are commercialized yearly in the US. The remaining 78 % go unlicensed. There is little doubt that the US technology transfer industry since Bayh-Dole has greatly enhanced new technology business formation, leveraging taxpayer-funded research and in some cases creating entire new industries i.e., biotech.

With the exception of France, most European and Asian countries have emulated the Bayh-Dole Act in their respective territories, giving the universities the ownership privilege and responsibility to commercialize new discoveries resulting from sponsored research. Although this has increased the supply of university discoveries in these territories, it has done little to increase demand or help build an effective market for university discoveries (Gross 2003).

The number of university discoveries and subsequent licenses is dependent on the research capacity of universities, namely their ability to develop new and unique IP. As a result, the best proxy for university inventiveness is the quality and number of patentable technologies produced. Historically, this has been related to the amount of sponsored research funding the institutions receive and the number of scientific, peer reviewed papers they publish. From 2001 to 2010, research expenditures have approximately doubled for the US universities, reaching $59 Billion USD in 2010 (Fig. 7.4).

About two-thirds of these investments in basic research have come from federally funded programs through the National Science Foundation, the National Institutes of Health and other government controlled research organizations. The percent of government contribution to research sponsorship has remained remarkably constant in the US during the last decade. The number of invention disclosures at these institutions has increased 39 % during the same period

	2001	2002	2003	2004	2005	2006	2007	2008	2009	2010
Total Research Expenditures ($ billions)	30	35	39	41	42	45	49	52	54	59
% Federal	64%	64%	66%	67%	67%	68%	65%	63%	62%	66%
% Industrial	8%	8%	7%	7%	7%	7%	7%	7%	7%	7%
Actual Federal ($ billions)	19	22	26	28	28	31	32	33	33	39
Actual Industrial ($ billions)	2	3	3	3	3	3	3	4	4	4

Fig. 7.4 Total research expenditures at U.S. Universities (2001–2010) (AUTM 2010a, b)

	2001	2002	2003	2004	2005	2006	2007	2008	2009	2010
Number of Respondents	169	188	198	198	191	189	193	189	181	183
Invention Disclosures Received	12,624	14,398	15,510	16,811	17,382	18,874	19,827	20,115	20,309	20,642
Disclosures per Respondent	75	77	78	85	91	100	103	106	112	113

Fig. 7.5 Invention disclosures from the U.S. Universities (2001–2010) (AUTM 2010a, b)

	2001	2002	2003	2004	2005	2006	2007	2008	2009	2010
Number of Respondents	168	189	198	192	178	172	194	191	181	183
New Patent Applications Filed	6,389	7,319	7,921	10,517	10,270	11,622	11,797	12,072	12,109	12,281
Total U.S. Patent Applications Filed	10,687	12,222	13,280	13,803	14,757	15,908	17,589	18,949	18,214	18,712
New U.S. Prov. Patent Applications Filed	N/A	N/A	N/A	6,192	6,640	7,856	8,328	8,865	8,364	8,653
New U.S. Utility Applications Filed	N/A	N/A	N/A	2,096	1,794	2,181	1,797	1,858	1,785	1,884
New Non U.S. Patent Applications Filed	N/A	N/A	N/A	1,277	1,102	1,403	1,070	848	1,322	1,116
U.S. Patents Issued	3,559	3,501	3,933	3,680	3,278	3,255	3,622	3,280	3,417	4,469

Fig. 7.6 Patent applications and issuances from the U.S. Universities (2001–2010) (AUTM 2010a, b)

(Fig. 7.5). The percent of invention disclosures that are filed as patent applications has grown from 85 to 90 %, and reached 18,712 patent applications in 2010 (Figs. 7.6 and 7.7).

Innovative technology is vital for the growth of companies, industries, and the economy. Writ large, the inexorable advance of technology is reflected in the patent activity of firms, industries, and countries. Patents, while not innovations themselves, serve as a proxy for invention and the seed corn for innovation and job creation. In 2006, more U.S. patents were granted than in any previous year (Perry 2010).

Although patent filing has continued at a good clip, the rate of issuance of patents in the US has slowed (Quinn 2011). There was a particularly large run-up

2010 Reporting Period	Respondents	Licenses/ Options Executed: Total	Licenses/ Options Executed: Startups	% of Total	Licenses/ Options Executed: Small Companies	% of total	Licenses/ Options Executed: Large Companies	% of Total
U.S. Universities	155	4,735*	830	18%	2,267	49%	1,562	34%
U.S. Hospitals & Research Institutions	27	621	78	13%	241	39%	302	49%
Technology Investment Firms	1	6	0	0%	0	0%	6	100%
All U.S. Respondents	183	5,362	908	17%	2,508	48%	1,870	36%

Fig. 7.7 Licenses executed by the U.S. Universities (AUTM 2010a, b)

in patent activity during the tech boom of the 1990s. This followed large inflows of capital into new technology companies that leveraged the Internet and related information technology.

However, as Seidel (2012) has made clear, innovation can never be fully described by the number of patents issued, as the go-to-market strategy is necessary to manifest the improvements in products and services.

After 1999, the number of patents filed increased faster than the number of issued patents. This is problematic, especially for SME's, which often rely of new patent issuances to raise equity financing. Nothaft and Michel (2010) believe that the patent office delay is due to a reduction in funding for the USPTO.

Nothaft and Michel (2010) estimated that approximately one million patent applications are currently awaiting examination by the USPTO and that 3–10 new jobs are created for every newly issued patent. If correct, approximately 2.5 million new jobs would be created in the US if the USPTO review backlog were resolved. The current backlog is approximately 3 years from date of filing. The usefulness of patents grows exponentially in the context of a well-financed corporate actor. Therefore, the need for linkage to the marketplace with the assistance of tech transfer organizations as well as the securing of financing to commercialize the discovery cannot be overlooked.

Even with 3 year of delays in patent prosecution times, about 30 % of patents applications are awarded as issued patents in the subsequent 3-year periods (Fig. 7.8).

Approximately 25 % of the university patents applied for in a given year are licensed or optioned to industry. With 48 % of these license or options going to small companies (Fig. 7.5), this underscores the importance of the availability of VC to create a successful technology transfer ecosystem. Companies of all sizes need new technologies but small companies (less than 500 employees) need access to VC to commercialize their technologies and build out their organizations. VC investors need the availability of exits (either IPOs or trade sales) to be encouraged to make these investments.

The economic value created for the U.S. universities as a result of this licensing activity was approximately $2.39 Billion USD in 2010, or 4.0 % of the $59 Billion USD invested in sponsored research at these institutions (Fig. 7.6). The knock-on economic and social benefits are much larger and more difficult to

	2001	2002	2003	2004	2005	2006	2007	2008	2009	2010
Number of U.S. Respondents	170	186	194	196	188	187	188	188	180	182
Total License* Income ($ millions)	$1,111	$1,304	$1,419	$1,474	$2,130	$2,173	$2,383	$3,444	$2,326	$2,396
Running Royalties ($ millions)	$825	$983	$1,126	$1,122	$1,139	$1,173	$1,938	$2,303	$1,618	$1,382
Cashed-In Equity ($ millions)	$104	$19	$39	$29	$43	$53	$46	$44	$24	$63
License Income of all other types ($ millions)	$123	$181	$157	$197	$758	$415	$216	$306	$282	$368
License Income Paid to Other Institutions ($ millions)	$82	$69	$108	$89	$85	$103	$97	$134	$173	$186
Number of U.S. Respondents	167	186	195	196	189	187	188	189	178	178
Licenses/Options Yielding Income	9,046	10,128	10,682	11,414	12,254	12,684	14,387	15,498	16,331	16,205

Fig. 7.8 Technology transfer income from the U.S. Universities (2001–2010) (AUTM 2010a, b)

measure. As an example, the market capitalization of Google (the recipient of a tech transfer from Stanford University for an improved search engine algorithm that started it all) is currently $193.7 Billion USD or about 4X the entire U.S. university research investment in 2010. Or perhaps even more important, the development of the PSA (prostate specific antigen) test for prostate cancer by the Roswell Park Cancer Institute, which was also a successful tech transfer that has saved tens of thousands of lives worldwide since its introduction.

Venture Capital

The mission of VC is to create a financial return from sagacious investments of capital and guidance into emerging, high potential businesses. What provides the high potential in these emerging companies is usually an amalgam of leadership talent and intellectual capital.

VC provides the essential financing that these companies need to commercialize their products and services, as emerging companies are usually too small to access the capital markets and too risky to qualify for commercial bank loans. In the US, VC has made possible the growth and development of two generations of firms, some of whom have become industry stalwart's (e.g., Apple), wealth for investors in these VC funds and according to the National VC Association (2012a, b), "12.1 million jobs and $3.1 Trillion in revenue in the United States in 2010," as a result of investing $39 Billion USD in 3,752 companies in 2011. Figures 7.9 and 7.10 show the number of deals and amounts of VC invested between 2003 and 2009 in the US

As a result of the global recession in 2008, there were approximately 1,000 fewer VC investments made in 2009 compared with the previous year and roughly a reduction of $10 Billion USD in the amount invested in 2009 in venture backed companies.

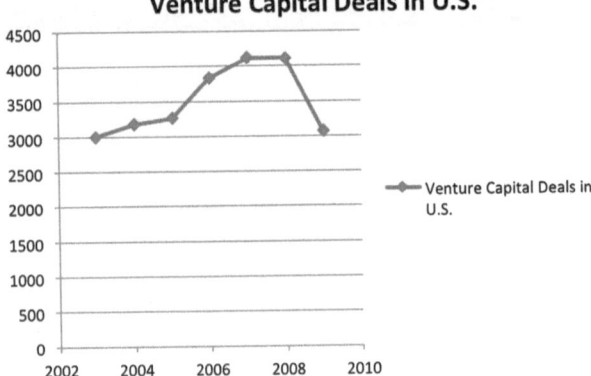

Fig. 7.9 Number of venture capital deals in the US. (2003–2009). National Venture Capital Association (2012a, b)

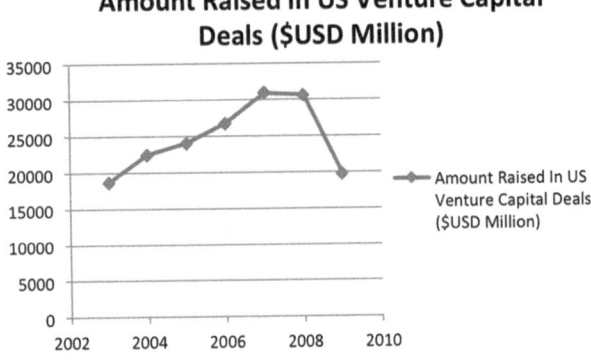

Fig. 7.10 Amount of venture capital invested in the US. (2003–2009). National Venture Capital Association (2012a, b)

The IPO Market

The IPO market is an essential source of financing for companies in all industries that have historical performance or compelling potential to create marketplace value for its customers and investors. Venture-backed companies usually seek exits in either IPOs or trade sales. As a result, VC exits and follow-on financings for the commercialization of new technologies require a functioning IPO and M&A marketplace. The number of IPOs and amount of capital raised in IPO financings is indicative of both the robustness of the capital markets and the overall growth of the economy as measure by GDP. From 2003 to 2010, there has been a reduction in the number of IPOs completed as well as the capital raised (Figs. 7.11 and 7.12). As a result of the recession of 2008, there were only 36 IPOs in the US that year, raising less than $30 Billion USD. The market has improved

Fig. 7.11 US IPOs by year
(E&Y 2012)

Fig. 7.12 US IPOs- money
raised by year (E&Y 2012)

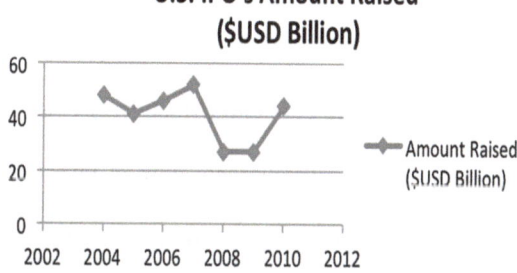

somewhat as of late with 163 IPOs consummated in 2011 raising $43.5 Billion
USD. While a good come back from the doldrums of 2008–2009 this is still far
behind the average number of IPOs between 1990 and 2000 which hovered in the
500+ transactions per year range.

For companies to grow and prosper, technological innovation is paramount.
Technologic innovation in companies relies on many factors principally the
availability of innovations, a capable team and availability of growth capital.
Universities are the largest source of innovative discoveries in the form of technologies available for license.

In the US, collectively, small firms add roughly three million jobs in their first
year of operations, a stark contrast to the one million jobs shed per year by larger
firms (Weitekamp and Pruitt 2010).

According to Levensohn (2010), for those firms that are able to access the
public equity markets approximately 90 % of job growth occurs post-IPO. Currently, this is not happening as vigorously as many would like, as a result of the
global recession, coupled with the unintended consequences of an overreaching
legislative framework, as well as technological advances that have dis-intermediated brokerage firms.

In the 1990s, the number of IPOs grew rapidly as investors were hungry to deploy
capital with early stage, technology enriched firms (Fig. 7.13). A period of job
creation, economic surpluses, and increased business process efficiency resulted.
These benefits came with the increased downside risk of tortious activities from

Fig. 7.13 China's
investment outpaces GDP
(Scissors 2011)

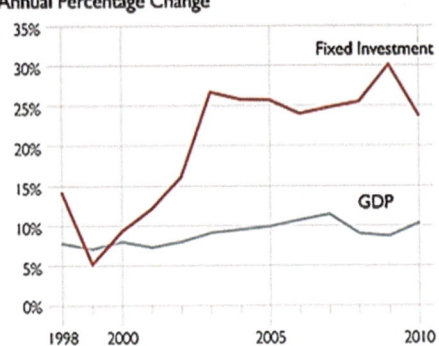

Source: *China Monthly Statistics,* Vol. 1 (1999)–Vol. 1 (2011), Beijing,
National Bureau of Statistics.

firms like ENRON and WorldCom, whose bankruptcies inspired numerous financial acts designed to stem future abuses. Regulation FD was enacted to separate investment bank research from trading activities and additional rules e.g., Sarbanes–Oxley, were promulgated to ensure better financial reporting and controls. Also, improved Internet technology allowed investors to buy and sell stocks directly, without the assistance of a broker. This reduced both the importance of stock brokering and the income such activities generated for the brokerage houses.

Stock quotes were decimalized from 1/16th of a dollar to 1/100 of a dollar (6.25–1 cent) to reduce bid-ask spreads. These narrower spreads resulted in reduced profits for market makers which led to reduced liquidity as there was less capital available to invest (Serchuk 2009). These factors reduced the market for new IPO's post 2000 (Ritholtz 2009). Small cap IPOs suffered the most (IPOs raising less than $50 Million). Many small cap banks closed-up shop, as they could no longer earn enough money by underwriting small offerings with their newly increased overhead and liabilities. On-line stock trading and REG FD further reduced their brokerage income. In short, the once robust IPO market was decimated, especially for the smaller offerings. Currently, only a handful of small cap investment banks are left in the US. (Borer 2012). As most new job growth is in small companies, the reduction in IPOs hit job creation and GDP growth hard. In 2011, there were less than 165 IPOs in the US, with an average size of $265 Million (The Business Review 2012), much less than the 520 per year needed to support 3 % GDP growth (Weil 2010).

The Chinese University Technology Transfer Market

Technology transfer requires the transfer of IPR. It is an ephemeral business that rests squarely on the foundation of IP law for its substance and transactions. Therefore, any discussion of tech transfer in China needs to begin with the current state of the IP laws and their intent (Thomson and Sigurdson 2008).

According to World Intellectual Property Organization (2007)

> The Decision on the Reform of Scientific and Technological Systems by the Central Committee of the Chinese Communist Party in 1985 marked this turning point in Chinese science and technology policy. This decision allowed universities to make their own decisions, based on the market situation, in organizing R&D programs and transferring technologies. In addition, the decision made it possible to provide incentives through more pay for more work. The role of the government changed from direct intervention and control to guidance and oversight, setting laws and regulations under which universities could decide on their own course of action.

It difficult to imagine, in consideration of the significant strides that have been made, that the Chinese patent office enacted the patent laws as recently as 1984 and subsequently amended them in 1992 and 2000, just 30 years in total (Ordish and Adcock 2008). According to the Patent office of China (SIPO 2012), one of their stated missions is to:

> With IP as a link among industry, academia and research community, improve the transferring and utilizing mechanisms for innovative results, promulgate policies that will promote IP transference and utilization, propel the translation of IP from laboratory to market as well as the commercialization and industrialization of IP.

This is consistent with the National Medium and Long-term Program for Science and Technology Development (STD 2006–2020) (Wu 2010). The goal of this program is to transform China from the world's factory to the world's innovator.

According to STD 2006–2020 (Wu 2010), China's science and technology focus will be to:

- "give priority to technological development in 11 major sectors such as energy, water resources and environmental protection in the coming 15 years
- further improve the national IPR system and strengthen the enforcement of IPR protection laws and regulations
- encourage enterprises to play the key role in innovation through their involvement in state projects and the provision of tax incentives and other financial support
- boost investment in science and technology; by 2020, China's research and development expenditures will account for about 2.5 % of the country's gross domestic product (GDP)
- by 2020, derive 60 % or more of its economic growth from technological progress; the numbers of patents granted to and total citations of journal articles by Chinese nationals are expected to be ranked among the top five in the world."

For China to achieve these heady goals, research and development activity (and funding) would have to increase exponentially. This is exactly what has happened (Fig. 7.11). Between 2005 and 2009, R&D expenditures have grown at an average real rate of 19.4 % (Wu 2010). From 1992 to 2008, R&D personnel increased from 670,000 to 1.9 million full-time workers (Wu 2010) and for the first time exceeding the number of US R&D employees, and R&D intensity (expenditure as a percent of GDP) has increased from 0.71 to 1.62 % from 1990 to 2009 (Wu 2010).

Country	R&D Expenditure (PPP$ billion in 2000 prices)	Business (%)	Government (%)	Others (%)	R&D Personnel (million people)
US	311.4	66.2	28.3	5.5	1.426
Japan	124.6	77.7	15.7	6.7	0.938
China	87.1	70.4	24.6	5.1	1.736
Germany	58.7	68.0	27.8	4.3	0.506
France	35.6	52.0	38.2	9.8	0.372
United Kingdom	33.3	46.5	30.0	23.4	0.349

Fig. 7.14 World's top five research and development spenders in 2007 (Wu 2010) [*PPP* purchasing power parity, *Note* Research and development personnel figures for the United States are 2006 data. *Sources* Research and development expenditure and personnel data are drawn from the OECD online database (<www.oecd.org>)]

China now ranks third amongst the world's leading R&D investors (Fig. 7.14). This expansion in R&D capacity has led to the dramatic increase in patent filings by universities (Fig. 7.15) and international patent filings (Fig. 7.16) and overall patent applications (Morrison 2011).

It is, however, a classic case of form over substance, with a few notable exceptions. Universities are ranked based on patent filings, an easier to measure metric than scientific discovery and highly important for academic advancement and receipt of government funding for research. Few of the Chinese universities have offices of IP management, and as a result there is a haphazard approach to IP protection, filing, and marketing (Orcutt and Shen 2010 and Feng (2009). Adam Smith's invisible hand seems to work equally well with new market entrants.

Yet, the numbers are impressive, if not the quality. According to Cyranoski (2010), China has the world's leading CAGR in new patent filings, and for the first time, has equaled the US in new patent filings (Morrison 2011).

A partial knock-on effect from the increase in patent activity has been the growth of technology transfers between 1999 and 2005. During this period, the number of technology licensing contracts with industry increased by 280 % to 842 licenses in 2005 generating $21.8 Million USD in licensing revenues.[1] By comparison in 2005, the US universities had completed 14, 284 license deals and generated license income of $2.4 Billion USD.

The growth of technology transfers in China has resulted from the precipitous increase in patent filings by universities. Patent filings alone, however, are not a sufficient proxy for technology transfer.

---

[1] Feng (2009).

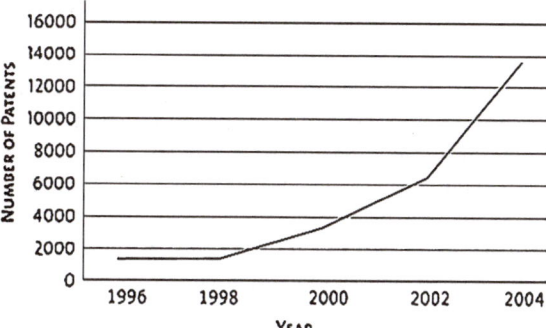

Fig. 7.15 Patent filings by Chinese universities (Guo 2007)

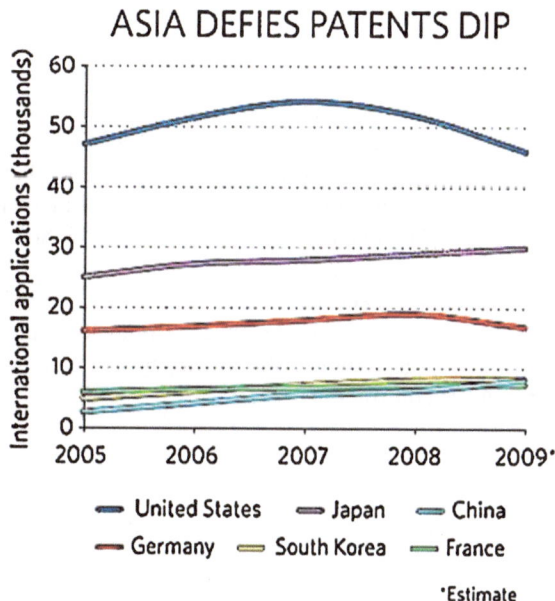

Fig. 7.16 World patent trends (international applications) (Cyranoski 2010)

For most Chinese universities, IP management essentially means making patent applications. Most Chinese universities do not have anyone specifically responsible for technology transfer. Without a TTO or anyone in charge of technology commercialization efforts, little effort is made to promote the actual transfer or commercialization of the resulting patents. In many cases, patents and commercially valuable research results are simply left on the shelf. Little is done to publicize them, making it hard for industry to learn about new technologies. Even when an entrepreneur might be informed and interested in licensing a patent, it is often unclear who in the university has the authority to negotiate.[2]

---

[2] According to Guo (2007).

The Technology Transfer Ecosystem

A robust IP framework is necessary for the commercialization of research discoveries. China established a patent office in 1980 and has made significant progress since the 1984 Patent Act in the filing and respect for IP. Currently, China has caught-up with the US on patent application filings (Fig. 7.28, Morrison 2011), although the quality of patents may be somewhat questionable. Most Chinese patents are design or utility patents (with 10-year terms) as opposed to the more fundamental innovation patents (20-year terms) which rely on basic research for their development.

According to Thomson Reuters (2011) recent report:

- "China has become the third-largest patent office in the world after the US and Japan by annual invention patent applications
- Published applications have increased by 16.7 % per year over the period from 171,000 in 2006 to nearly 314,000 in 2010.
- China's global ranking based on citations in international science papers has moved from 13th in 2006 to 8th in 2010.
- China now ranks 29th as measured by the Global Innovation Index and is the only developing country among the top 30 innovators.
- In 2008, China invested 457 billion Yuan (US$65.8 Billion) in R&D, or 1.52 % of its increasing GDP.
- Significant recent technological achievements include the opening of the world's longest sea bridge at 26.4 miles long and the ongoing development of world's largest high-speed railway system with around 10,500 miles completed or under construction."

China's Venture Capital Marketplace

During the past 25 years, China has developed a very active VC ecosystem providing capital and guidance to emerging companies (Zhang et al. 2009 and Orcutt and Shen 2010). In 1984, VC industry was legalized in China and subsequently modernized in 1998. Convertible preferred stock, a staple of VC investing has only been allowed in China since 2006 (Orcutt and Shen 2010). From an almost standing start in 1997, only seven venture investments were made in China (Orcutt and Shen 2010). Early venture investments in China focused on infrastructure and real estate projects due to their lower risk and as a result technology investments were limited. The bankruptcy of China's first VC company, a state-owned company (China New Technology Venture Capital Company Ltd.) did not help matters, but it did serve as a good example of the adverse selection that results when state goals trump investment returns. In 1998, the Proposal for Developing a VC Industry in China [known as the No. 1 Proposal- (Orcutt and Shen 2010)] received

Fig. 7.17 Number of venture capital deals in China (Spender 2010)

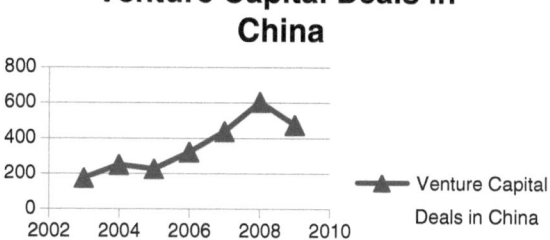

Fig. 7.18 Amount of venture capital raised in China (Spender 2010)

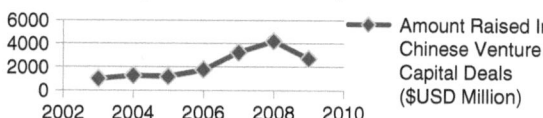

wide-scale support (following the US model for VC's) which helped to launch the VC industry in China. In 2006, the Provisional Measures for the Administration of VC were the first national statue to govern domestic VC investing in China (Orcutt and Shen 2010). From less than 200 investments in 2003, nearly 500 VC investments were made in 2009 (Fig. 7.17) resulting in $2.5 Billion USD of investment in 2009 (Fig. 7.18).

China's IPO Market Place

To encourage VC investments, China has developed an active public equity marketplace in both Shanghai and Shenzhen, in addition to the internationally recognized Hong Kong Stock Exchange. The preferred exits for VC investments in China are IPOs (as opposed to trade sales) and as a result, in addition to the major trading markets China has also introduced alternative or Growth Enterprise Markets (GEMs) with less strict listing requirements. Similar to the AIM market on the LSE in the UK or either the AMEX-NYSE or the NASDAQ Capital Market in New York; GEMs are designed to make it easier for emerging companies to raise capital and provide liquidity for VC investors. ChiNext, China's first GEM was launched by the Shenzhen Stock Exchange in 2009 (Orcutt and Shen 2010). IPO proceeds on CHiNext reached $31.1 Billion USD in 2011 with the total market capitalization of listed companies on this exchange reaching $118 Billion USD in 2011 (Shenzhen Stock Exchange 2012).

Fig. 7.19 Chinese IPOs by year (number of deals) (E&Y 2012)

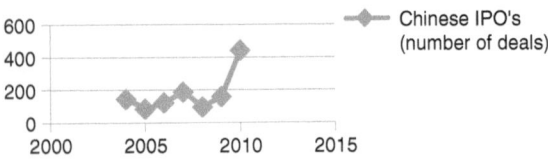

Fig. 7.20 Chinese IPOs by year (capital raised) (E&Y 2012)

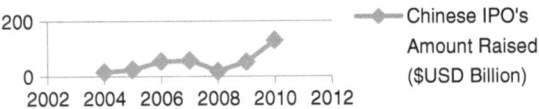

From 2004 to 2010, the number of IPOs in China tripled to 440+ deals raising approximately $130 Billion USD (Figs. 7.19 and 7.20) (E&Y 2012)

Recently, however, China has taken second place after the US on the completion of IPOs in Q1 2012 (Cowan 2012). This is a turn-around of a 3year trend with the seemingly endless supply of Chinese IPOs. In Q1 2012, Chinese IPOs raised US$10.3 Billion versus US$23.1 Billion in Q1 2011. The US IPOs raised USD $11.9 Billion in Q1 2012. The performance of recent Chinese IPOs has not been good and as a result the capital markets appetite for them has been tempered.

Of late, Chinese IPOs have become more difficult to sell in the US due to accounting and corporate governance lapses and several high profile failures. In China, many of the new offerings have also not performed well in the aftermarket, creating uncertainty and reduced investor interest there as well, although the market is still moving forward.

University Technology Transfer Marketplace: Growth Over the Next Decade

University-based scientists and engineers are the producers of the intellectual capital that drive the technology transfer and spinout companies. While the US has the largest and most developed university graduate science and engineering education system in the world, more than 50 % of engineering Ph. D. degrees awarded annually in the US are earned by foreign nationals (Wadwha et al. 2007); the majority of which are Chinese and Indian students. Approximately 40,300 master's students and 10,570 Ph.D. students from China were studying engineering in the US in 2009 (Burrelli 2010) and approximately 30 % of all Chinese students

Fig. 7.21 Ten-year trend in engineering and technology master's degrees in the US and China (edited after Wadwha et al. 2007). *Note Hashed line* represents outliers removed by Wadwha et al. (2007)

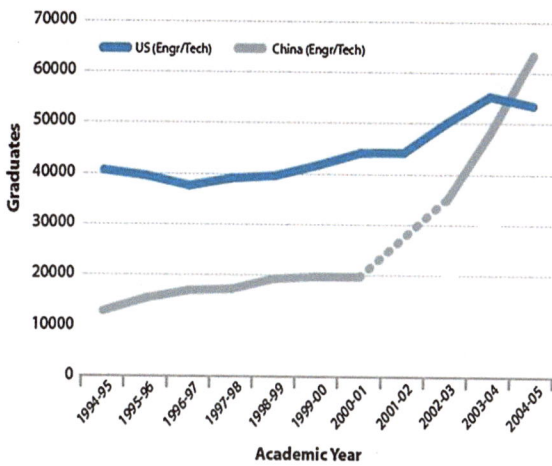

Fig. 7.22 Ten-year trend in engineering and technology Ph. D. degrees in the US and China (edited after Wadwha et al. 2007). *Note Hashed line* represents outliers removed by Wadwha et al. (2007)

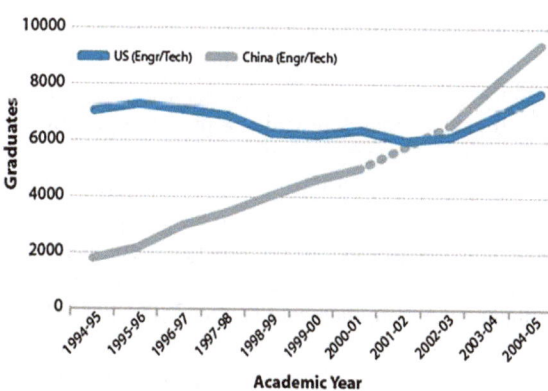

studying abroad returned home following their university education. By 2005, the number of master and doctoral degree graduates in engineering in China (Figs. 7.21 and 7.22) has surpassed the number of graduates produced by the US and the trend is accelerating. This is a remarkable achievement when one considers that the number of Chinese masters engineering graduates were less than half of those produced in the US in 2000.

To potentially help address the departure of talented foreign born graduate students, a recent law has been proposed in the US called the Start-up Visa Act of 2011, as an amendment to the immigration law of the US, Its goal is to create a new visa category (EB-6) specifically designed to permit nonUS citizen entrepreneurs who have raised at least $100,000 from American investors, in start-up capital, to obtain a working visa and ultimately a green card, if the business creates at least five new jobs and raises either $500,000 or has sales of $500,000 in the subsequent two-year period from founding (Wikipedia 2012a, b). In addition, the requirements are somewhat more relaxed for individuals who have graduate degrees in science and engineering; if they have an annual income of $30,000 or assets of at least $60,000

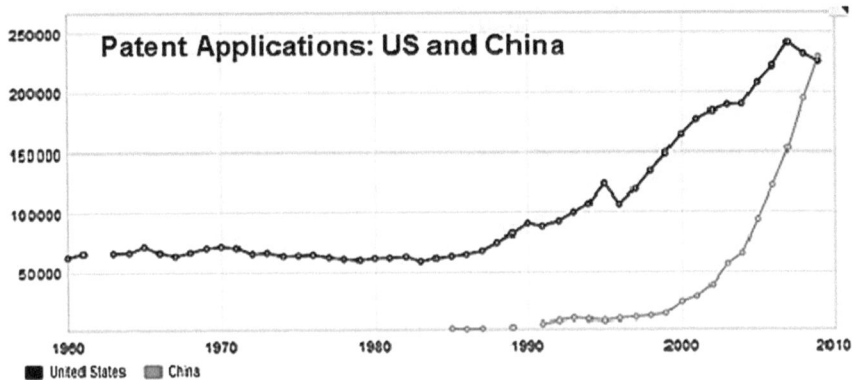

Fig. 7.23 China has caught up to the US in patent applications in 2010 (Morrison 2011)

and a qualified US investor has agreed to invest at least $20,000 in their new business, they will be eligible for the visa (The Start-up Business Act 2011).

This increase in the population of scientists and engineers combined with the Chinese Government's emphasis on patent filing and innovation has resulted in significant increases in China's patent application capability (Fig. 7.23).

Innovation patents are the seed corn for the development of technology-based companies. University patents play a disproportionate role here as large corporate inventions are usually captive to the companies that create them and as a result not available for transfer to the emerging companies. Although it is called R&D on corporate accounts, most companies, even large ones pursue "D" in their quest to enhance their products and services. The big "R" research is most often relegated to university and federal research centers where scientists have the mandate (some might say luxury) to investigate fundamental knowledge quests.

In an attempt to stake their claims on the IP landscape, university scientists and engineers in both the US and China are filing record numbers of patents, which is hugely positive but not sufficient to catalyze the low probability matching process that companies must go through to create innovation.

Nevertheless, the projected increase in patent applications for key global actors puts China on a fast paced, world leading growth trajectory and has in fact as of 2013 surpassed other nations as projected (Fig. 7.24).

A measure of the quality of the intellectual capital being produced by scientists, as opposed to the quantity of participants, can be gaged from the science and technical publications in peer reviewed journals. According to the World Bank (2012), the number of scientific and technical journal articles produced by China between 2007 and 2009 increased by 17,208–74,019 papers, whereas publications by the US authors, while almost 3X as frequent as Chinese authored papers (208,601 papers in 2009), were slightly down during the same period (Fig. 7.25). Projecting the trend forward toward 2020, China is on track to match the output of the US science and technology publications by 2017 (Fig. 7.26, extrapolation using exponential curve fitting) and then surpass the US in this important measure by 2020.

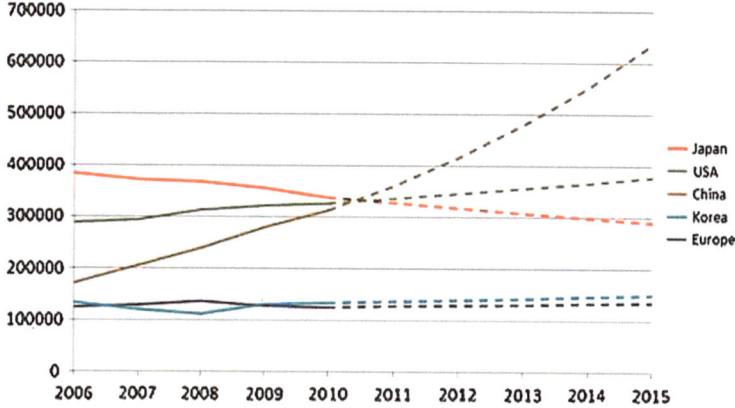

Fig. 7.24 China's patent growth trajectory (Thomson Reuters 2011)

	2007	2008	2009
China	56,811	65,301	74,019
United States	209,898	212,883	208,601

Fig. 7.25 Scientific and technical journal articles (The World Bank 2012). *Note* The numbers of scientific and engineering articles published in the following fields: physics, biology, chemistry, mathematics, clinical medicine, biomedical research, engineering and technology, and earth and space sciences. Data extracted from National Science Foundation, Science and Engineering Indicators

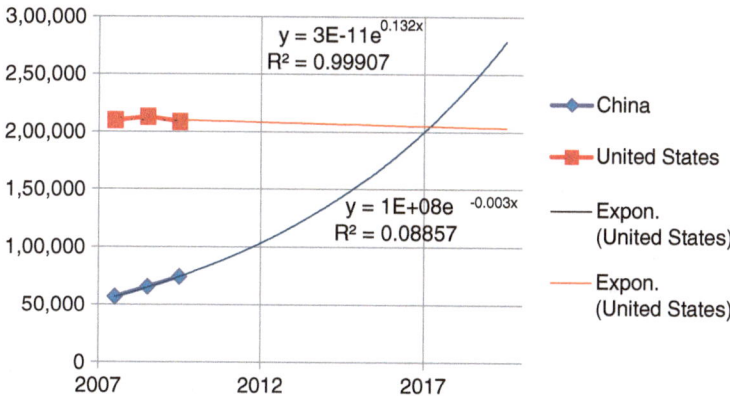

Fig. 7.26 Projected scientific and technical journal articles to 2020 (from World Bank Data 2012) (extrapolations by the author are rough estimates of what the relative US versus China journal article production may resemble in 17 years' time)

Implications for Global Markets

Long-term with regard to the impact on the global market, perhaps the most telling number is the relative R&D spend by the US and China, as together, both countries represent approximately 46.9 % of total global GDP (Fig. 7.27, R&D MAG.com 2011), with the US representing 34 % and China 12.9 %, respectively. The R&D investment by the US, combined with its available supply of scientists and engineers to both create and improve new discoveries, has positioned it as the leading producer of intellectual capital. These innovations have resulted in improvements in products and services across industries, along with significant improvements in worker productivity, even when taking into account significant job losses since 2008.

However, China's IPO market has achieved lift-off and is now responsible for almost half of the world's IPO's (Fig. 7.28), raising $6.2 Billion USD in Q1 2012 across 60 IPOs. This represents the confluence of pent-up demand for equities both as an asset class as well as means to exit an equally robust VC investment trajectory.

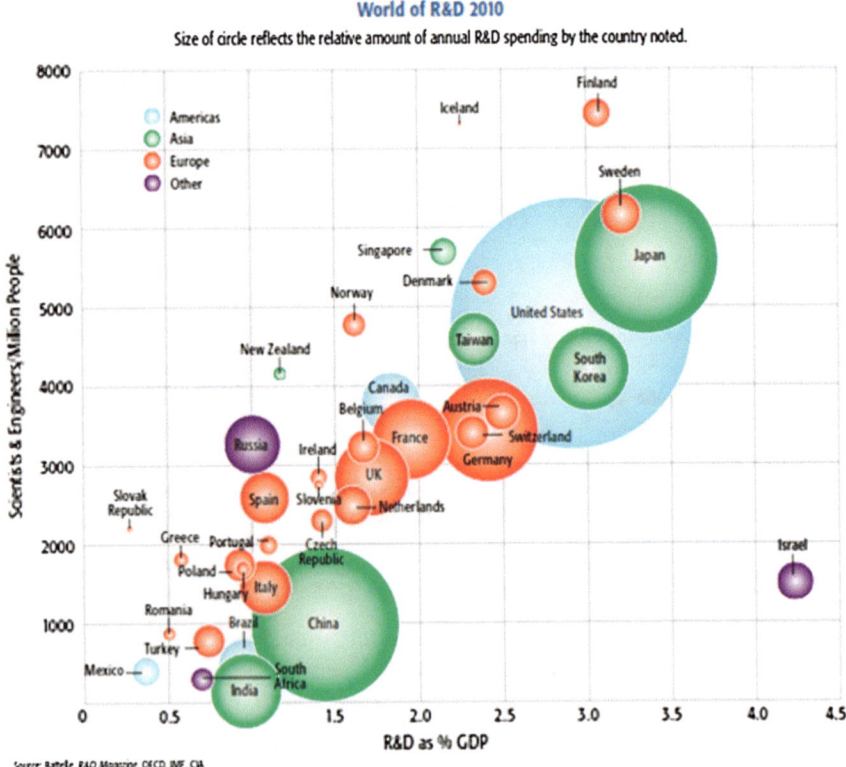

Fig. 7.27 Global R&D spending in 2011 (Clancy 2011) *Source*: Battelle/R&D Magazine 2012 Global R&D Funding Forecast

Table 1 Comparison of Global IPOs in Q1'12 by Enterprise Type					
Enterprise Type	Financing Amt. (US$M)	% of Total	No. of IPOs	% of Total	Average Financing Amt. (US$M)
Chinese Enterprises	6,206.58	47.8%	60	50.8%	103.44
Foreign Enterprises	6,775.36	52.2%	58	49.2%	116.82
Total	12,981.94	100.0%	118	100.0%	110.02

Source: Zero2IPO Research Center, Mar. 2012   www.zdbchina.com

Fig. 7.28 Global IPO marketplace, Q1 2012 (Zero2IPO 2012)

When considering the amount of VC deployed, both in terms of the number of deals and the dollar amount invested, the US market is in steep decline (Fig. 7.29), and the exits, however, for these VC investments look far less certain in the US than they do for China.

To help address the downturn in IPO's which has been going on since 200, the Jobs Act was recently signed into law (April 7, 2012) by President Obama, relaxing Sarbanes–Oxley requirements for small public companies (less than $1 Billion in market cap) for the first 5 years post IPO. Specifically, these companies won't need to audit their internal controls, although there will be a requirement that they have the internal controls in place. Additionally, the Act allows small private companies to sell shares to up to 499 nonaccredited investors with less than $100k in yearly income (they can invest up to 5 % of their income). The ACT allows small companies to raise up to $1 Million without SEC registration. This important legislation has the potential to unblock equity financing for small companies while ushering in a new industry of crowd source financing. Uncharacteristically, both the Senate and the U.S. Congress have supported the ACT, recognizing the connectivity between equity financing, the growth of small companies and the creation of jobs in the U.S. Whether or not it is sufficient to address the downturn in the U.S. IPO market remains to be seen, but it is unlikely.

The current IPO situation in China is strong (Fig. 7.30) and projected to be quite robust with an estimated through 2020 if the trend continues. This should well service its projected VC industry with exits, which are projected to invest an estimated $40 Billion USD in 2020.[3]

From the current data, it is difficult to extrapolate 10+ years into the future but the best guess is that at the current rate of technology transfers the US will still dominate the industry in both quantities of transactions which are linearly estimated to grow to 25,000 and $5.5 Billion USD in licensing revenues (Fig. 7.31) versus 7,000 transfers and $93 Million USD in licensing income for China. Overall, this would imply a 2.3X increase in the size of the current market for tech transfers (based on linearly extrapolated licensing income) for technology transfers over the next 10 years, for China and the U.S.

---

[3] Gross (2012).

	2008	**2009**	**2010**
Number of deals	37	67 (▲81%1)	163 (▲143%2)
Capital raised (US$)	$26.8b	$27.3b (▲2%1)	$43.5b (▲60%2)
Average deal size (US$)	$724.9m3	$406.9m	$267.1m
PE-backed IPOs (number of deals, capital raised)	8 deals, $2.5b	28 deals, $9.0b	84 deals, $15.2b
Top sectors (number of deals)	Energy (8)	High technology	High technology
	Financials (5)	(12)	(35)
	Health care (5)	Health care (9)	Health care (21)
	Industrials (5)	Real estate (9)	Financials (20)
	High technology	Industrials (8)	Industrials (17)
	(4)	Financials (5)	Energy (14)
Top sectors (capital raised)	Financials	Financials ($10.5b)	Industrials ($22.0b)
	($20.0b)	High technology	High technology
	Energy ($2.7b)	($3.2b)	($4.9b)
	Materials ($1.3b)	Real estate ($2.9b)	Financials ($4.1b)
	Industrials	Health care ($2.2b)	Energy ($3.5b)
	($0.8b)	Energy ($1.4b)	Real estate ($2.0b)
	High technology		
	($0.6b)		
Stock exchanges			
NYSE	17 deals,	35 deals, $19.1b	82 deals, $34.7b
NASDAQ	$25.1b1	30 deals, $8.1b	76 deals, $8.7b
AMEX	18 deals, $1.7b	1 deal, $2m	5deals, $94.5m
(number of deals, capital raised)	1 deal, $3m		

2 Percentage change from 2009 to 2010
3 Includes Visa's $19.7b IPO (largest US IPO ever). The average deal size excluding Visa is US$199.2 million
Source: Dealogic, Thomson Financial, Ernst &Young Percentage change from 2008 to 2009

Fig. 7.29 US IPO statistics [E&Y 2012 (http://www.ey.com/GL/en/Services/Strategic-Growth-Markets/Global-IPO-trends-2011—United-States, Accessed 29 May 2012)]

	2008	2009	2010
Number of deals	97	159 (▲64%[1])	440 (▲177%[2])
Capital raised (US$)	$17.5b	$51.5b (▲194%[1])	$129.8b (▲152%[2])
Average deal size (US$)	$180.4m	$324.1m	$295.1m
Top five sectors (number of deals)	Materials (26) Industrials (24) Consumer staples (8) Retail (8) High technology (7)	Industrials (34) Materials (22) Consumer staples (19) High technology (18) Consumer products (15)	Industrials (103) Materials (97) High technology (70) Consumer staples (44) Health care (28)
Top five sectors (capital raised)	Industrials ($8.8b) Materials ($3.0b) Consumer staples ($1.8b) Retail ($1.0b) Energy ($0.8b)	Industrials ($19.7b) Materials ($5.4b) Real estate ($5.2b) Media & entertainment ($4.8b) Energy ($3.5b)	Financials ($51.1b) Industrials ($20.1b) Materials ($18.5b) High technology ($10.6b) Health care ($6.1b)
Stock exchanges: Hong Kong Shanghai Shenzhen – SME Shenzhen – ChiNext (number of deals, capital raised)	24 deals, $4.8b 3 deals, $8.5b 69 deals, $4.1b N/A	56 deals, $21.9b 8 deals, $20.4b 54 deals, $6.2b 36 deals, $3.0b	87 deals, $57.4b 26 deals, $27.9b 205 deals, $30.2b 116 deals, $14.1b

[1]Percentage change from 2008 to 2009
[2]Percentage change from 2009 to 2010
Source: Dealogic, Thomson Financial, Ernst & Young

Fig. 7.30 China IPO statistics (http://www.ey.com/GL/en/Services/Strategic-Growth-Markets/Global-IPO-trends-2011—United-States, Accessed 29 May 2012)

Fig. 7.31 Projected US and China technology transfers (number of transfers/year) (AUTM 2010 and Feng 2009) (extrapolations by the author are just rough estimates of what the relative number of the US versus China technology transfers may resemble in 17 years' time)

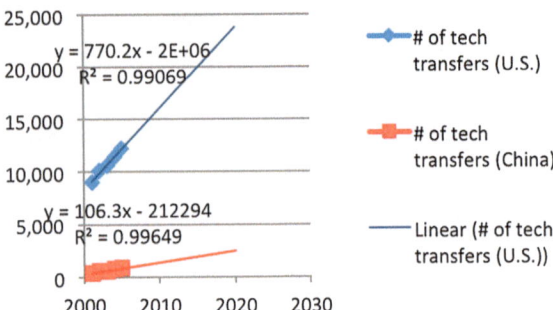

US technology transfer should continue to be robust in 10 years' time; however, a large number of the technologies are likely to be exported to China to take advantage of the robust and growing corporate, VC and IPO marketplace; especially considering the projected fall off in the US IPO market. To achieve these cross-border technology transfers would require an increase in the current skill set of university tech transfer offices or require the development of a class of tech transfer intermediaries that can efficiently bring cross-border purchasers and providers of technology licenses together with a minimum of friction and expense.

There is of course the possibility that China's growth will slow down appreciably or even stall, resulting in reduced amounts of VC investments and IPO's, not unlike what happened in Japan between 1990 and 2012 (Fig. 7.32) or the US in 2000 and 2006. Through perhaps the occurrence of a black swan or other unlikely event, the effect on technology transfer in China would nevertheless be modest, as most of the new technologies required to invigorate, what will then be the world's largest economy, can hardly be supplied by China's indigenous R&D efforts over the next 10 years. In spite of the impressive gains in the number of graduating scientists and engineers, the stock of knowledge, the number of patent applications, published scientific papers, and rapidly emerging equity markets. China will still need to import large quantities of new discoveries to exogenously enhance its R&D, increase its Total Factor Productivity and maintain an innovative corporate base for continued GDP expansion and employment. This bodes well for the global tech transfer markets as foreign firms will play a dominant role in supplying technology transfer in China over the next 10 years; especially the US and European organizations that have strong indigenous R&D capabilities with their excellent research universities, global access to new technologies and significant experience in technology transfer.

Generically, these firms will be of two types, either technology commercialization offices of foreign (nonChinese) research institutions or intermediaries that have strong access to innovations worldwide, coupled with the knowledge required for effectuating technology transfers.

The technology transfer market and organizational factors affecting the success of companies in both the US and China are inextricably interwoven. A deficiency in one country could have a knock-on effect of impeding GDP growth and

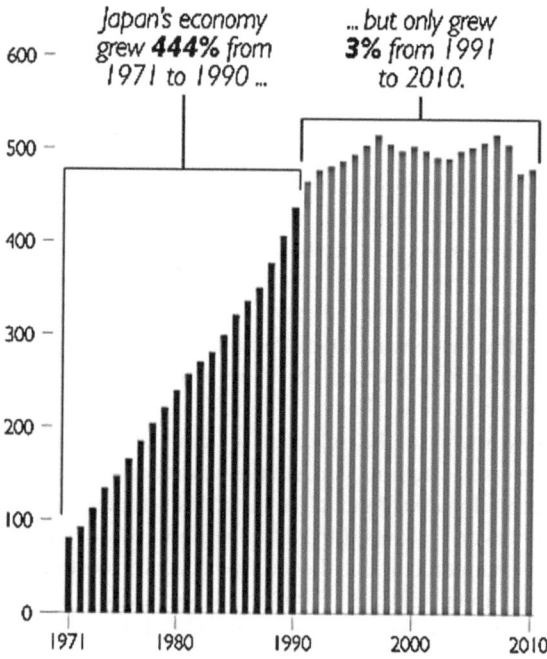

Fig. 7.32 The slowing of Japan's growth (Scissors 2011)

employment in both countries. For many companies in the US and China, universities are the best source for externally developed, potentially disruptive, new discoveries. In fact for the US, technology transfer should become even more important in the coming decade, a growth industry, as it seeks to monetize its technology development prowess and increase its exports of high technology products.

Problems with the Nascent Technology Transfer Market

University technology transfer is still emerging as a business sector. The majority of the actors in the tech transfer marketplace can be subsumed under the invisible college network strategy (Ventresca 2012). In this network model, universities are the suppliers of the new technologies and firms are the buyers. The market is delineated by the technology push model and as a result is much less efficient than it could be, as evidenced by the relatively few transfers being consummated, in all jurisdictions, compared with the available supply of new technologies (approximately 80 % of the US university discoveries go un-licensed each year).

The majority of the existing tech transfer firms have immutably strong ties with their home institutions that normally preclude them from sourcing discoveries from other institutions that might be more helpful to their clients. The holes in the university networks are the weak ties with the client companies (unless they are spin-outs). Stand-alone consultancies do not have this problem, as they are technology agnostic and usually have weak ties with their suppliers and strong ties with their customers; both of which are a virtue in the tech transfer marketplace. However, these firms have the disadvantage of not owning by fiat the technologies that university tech transfer companies have, and therefore must manage a discovery network of universities alongside of and on behalf of their customers. Gross (2003) has described a market driven technology transfer model that begins with the company that needs the technology and has strong ties with the technology acquirer (its customer) but little else. The firm is market driven, consistent with its technology needs and customers.

The customer, for example a Chinese company can readily leverage the tech transfer firm's weak ties to its university network to source the technologies it needs to grow.[4] For tech transfer firms, weak ties are preferred to strong ties as they increase the selection of available technologies and in doing so reduce adverse selection. This simplistic process basically reverses the normal university technology push model, in favor of customer centric, market pull. A key weakness, however, remains. Innovation requires more than technology, it also needs capital to commercialize it; hence, the need for customers to develop connectivity with the capital markets. For China over the next 10 years this should not represent a problem, as the recent surge in VC and the robust IPO marketplace should make it readily available, to those companies with proprietary innovations and capable management.

In the US, industrial companies spend about 3.5 % of revenues on R&D, while pharmaceuticals like Merck & Company spend 14 %, computer manufacturing companies 7 %, and biotech's like Allergan as much as 43 % of revenues on R&D[5]. The direct sponsored research costs to develop a single new technology, from conception to patent filing, at US universities is approximately $4.812 Million[6]. This estimate should be viewed as ultra-conservative, e.g., tip of the iceberg, as it does not include the infrastructure costs such as laboratories, computing facilities, fixed base salaries of researchers, physical facilities, patent expenses, and university technology management expenses among others. Given the steep cost for developing a new technology and the high failure rate, the logical e.g., cost-effective solution is for companies to embrace open innovation and

---

[4] Granovetter (1973).

[5] Wikipedia (2012a, b).

[6] Imputed from AUTM (2010).

exogenously source technologies from amongst those already developed using taxpayer-funded research at major universities.

This certainly is the case for both US and Chinese companies. However, companies in both countries will need to embrace technology transfer more aggressively than they have previously, to reap this advantage. For technology transfer to become main stream over the next decade, the a priori bias against technologies "not invented here" will have to be laid to rest and financial incentives for their commercialization must be implemented world-wide. Additionally, and perhaps more importantly, technology transfer allows companies to systematically embrace open innovation and source technologies simultaneously from multiple institutions, piecing together solutions from the virtual "new idea factory"[7] consisting of the best research universities in the world. Currently, there are approximately 3,290 research institutions in 160 countries that produce 80 % of the world's top scientific research. The breakdown by country is listed in Fig. 7.33. The US and China account for roughly 26 % of the leading research institutions producing peer reviewed, published scientific research. Finding the right new technology to invigorate a company's products requires the continuous evaluation of the fruits of research from all of these idea factories. If it is a worthwhile technology, it matter little where the idea was developed.

This is a high potential strategy to improve the productivity of companies in all jurisdictions, with potential knock-on benefits of enhanced GDP and employment. As IBM's recent CEO study has demonstrated (Fig. 7.34), having the "right technology" has presciently become the major concern of company leaders worldwide.

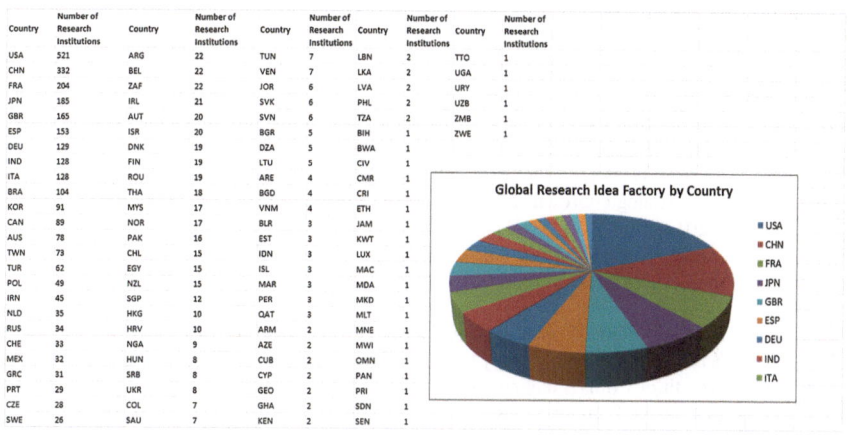

Fig. 7.33 The global research idea factory: # of institutions by country (SIR World Report 2012)

---

[7] Gross (2000).

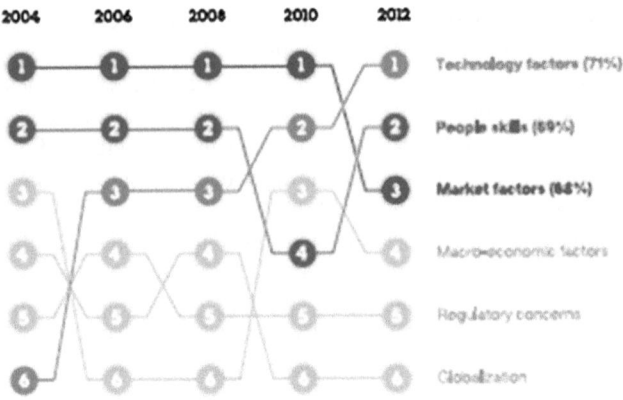

Fig. 7.34 External forces that can impact an organization over the next 3–5 years (from IBM's 2012 CEO study of 1,709 executives in 64 countries)

References

AUTM U.S.: Licensing activity survey: FY2010. In: Richard Kordal, Arjun Sanga, and Paul Hippeneyer editors with research assistance by Chrys Gwellem. Association of University Technology Managers. 2010a. AUMT.net, Accessed 1 May 2012.

AUTM U.S.: Licensing activity survey highlights: FY2010. http://www.autm.net/AM/Template.cfm?Section=FY_2010_Licensing_Survey&Template=/CM/ContentDisplay.cfm&ContentID=6874. 2010b. Accessed 29 Jan 2012.

Borer J. Personal communication. New York: Rodman and Renshaw; 2012.

Burrelli J. Foreign science and engineering students in the united states, NSF 10-324. 2010. http://www.nsf.gov/statistics/infbrief/nsf10324/. Accessed 5 May 2012.

Chesbrough H. Open innovation: the new imperative for creating and profiting from technology. Boston: HBS Press; 2003.

Clancy H. United States will lead 2011 R&D funding, China now no. 2. 3 Jan 2011.

Cowan L. China slips to no 2 for IPOs raised, behind U.S. www.dowjones.com. 8 May 2012. Accessed 1 June 2012.

Cyranoski D. China's patents push: Asia defies patent-filing downturn as global economy slips, Published online 15 Feb 2010, Nature. doi:10.1038/news.2010.72.

E&Y. From data from Dealogic, Thomson Financial, Ernst & Young. 2012. http://www.ey.com/GL/en/Services/Strategic-Growth-Markets/Global-IPO-trends-2011—United-States. Accessed 29 May 2012.

Feng TM. Technology transfer from university to industry: insight into university technology transfer in the Chinese national innovation system. London: Adonis & Abbey Publishers Ltd.; 2009.

Granovetter M. The strength of weak ties. Am J Sociol. 1973;78:1360–80.

Gross C. U2B: a new model for technology transfer. BioEntrepreneur. 2003;21. Chapter 7, p 87–90.

Gross C. The growth of China's technology transfer industry over the next decade: implications for global markets. J Technol Transf. Springer Netherlands. 2012. doi:10.1007/s10961-012-9263-x.

Gross C, Reischl U, Abercrombie P. The new idea factory. Columbus: Battelle Press; 2000.

Guo H. Chapter no. 17.9, IP management at Chinese Universities, Jones day, China, IP handbook of best practices. 2007. http://www.iphandbook.org/. Accessed 29 May 2012.

IBM Global Business Services: Leading through connections, New York; May 2012.

Levensohn P. http://www.pascalsview.com/pascalsview/2010/08/connecting-the-dots-how-new-job-creation-ipo%E2%80%99s-and-venture-capital-in-america-are-intimately-linked.html (2010). Accessed 4 Mar 2012.

Morrison M. America's perfect storms: part III—Economic challenges in a global society. 15 Sept 2011. http://www.decisionsonevidence.com/2011/09/america%E2%80%99s-perfect-storms-part-iii-%E2%80%93-economic-challenges-in-a-global-society/. Accessed 27 May 2012.

National Venture Capital Association. Year book 2012. Arlington, Virginia: Thomson Reuters; 2012a.

National Venture Capital Association. Web page "About Us," Arlington, Virginia. 2012b. http://www.nvca.org/index.php?option=com_content&view=article&id=339&Itemid=653. Accessed 5 June 2012.

Nothaft H, Michel P. New York times opinion article, inventing our way out of joblessness. 5 Aug 2010.

Orcutt J, Shen H. Shaping China's innovation future: university technology transfer in transition. Massachusetts: Edward Elgar Publishing, Inc.; 2010.

Ordish R, Adcock A. China intellectual property challenges and solutions. London: Wiley (Asia); 2008.

Palfrey J. Intellectual property strategy. Massachusetts: MIT Press; 2012.

Perry M. U.S. patent activity continues to grow, blog for economics and finance. 28 Nov 2010. http://mjperry.blogspot.com/2010/11/american-exceptionalism-continues-to.html. Accessed 4 Mar 2012.

Quinn G. IPWatchdog, Inc., Posted: Jan 2, 2011. http://www.ipwatchdog.com/2011/01/02/why-patents-matter-job-creation-economic-growth/id=14170/. Accessed 3 Mar 2012.

Ritholtz B. The big picture. 30 Jan 2009, 11:15 A.M. http://www.ritholtz.com/blog/2009/01/historical-chart-of-initial-public-offerings/. Accessed 1 Mar 2012.

Scissors D. The heritage foundation. 14 Apr 2011. http://www.heritage.org/research/reports/2011/04/the-united-states-vs-china-which-economy-is-bigger-which-is-better. Accessed 29 May 2012.

Seidel V. Handout for strategy and innovation course, models of innovation in technology based markets, EMBA8. Said Business School, Oxford University; 2012, p. 1, slides 4,6.

Serchuk D. Decimalization and its discontents, Forbes.com, 10 Mar 2009. http://www.forbes.com/2009/03/09/decimalization-uptick-rule-intelligent-investing-volatility.html. Accessed 3 Mar 2012.

Shenzhen Stock Exchange: http://www.szse.cn/main/en/ChiNext/. Accessed May 2012.

SIR World Report 2012: Global Ranking, data edited from SCOPUS. http://www.info.sciverse.com/scopus/ and SCIMAGO Research Group, http://www.scimagolab.com/.

Spender T. High risk, high impact, May 2010, Issue 40. http://www.scribd.com/doc/30688652/Venture-Capital-in-China. Accessed 2 June 2012. Data source, Zero2IPO.

State Intellectual Property Office of the PRC (SIPO): http://english.sipo.gov.cn/laws/. Accessed 29 May 2012.

The Business Review, March 6th, 2012 (quoting PWC, LLC), http://www.bizjournals.com/albany/morning_call/2012/03/ipos-were-down-20-percent-in-2011.html.

Thomson Reuters: Chinese patenting: report on the current state of innovation in China. http://blog.thomsonreuters.com/index.php/category/events/knowledgeexchange/ (2011). Accessed 27 May 2012.

Thomson E, Sigurdson J. China's science and technology sector and the forces of globalisation. Singapore: World Scientific Publishing; 2008.

The Start-up Business Act of 2011. http://www.feld.com/wp/archives/2011/03/the-startup-visa-act-of-2011.html. Accessed 5 June 2012.

The World Bank: http://data.worldbank.org/indicator/IP.JRN.ARTC.SC. Accessed 31 May 2012.

U.S. Congress: Office of Technology Assessment, technology transfer to China, OTA-ISC-340. Washington, DC: U.S. Government Printing Office; 1987.

Ventresca M. Handout #4, p20 & Handout #6, p7, for strategy and innovation course, EMBA8. Said Business School, Oxford University; 2012.

Wadwha V, Gereffi G, Rissing B, Ong R. Where the engineers are, issues in science and technology. Austin: University of Texas; 2007. http://www.issues.org/23.3/wadhwa.html. Accessed 2 June 2012.

Weild D, Kim E. Market structure is causing the IPO crisis and more. Londan: Grant Thornton, LLP; 2010.

Weitekamp R, Pruitt B. Job Growth in U.S. Driven Entirely by Startups, Kauffman Foundation Study, http://www.kauffman.org/newsroom/u-s-job-growth-driven-entirely-by-startups.aspx. July 7, 2010.

Wikipedia: Comparison between U.S. and countries nominal GDP. 2012a. http://en.wikipedia.org/wiki/Comparison_between_U.S._states_and_countries_nominal_GDP. Accessed 29 May 2012.

Wikipedia: Start-up visa. 2012b. http://en.wikipedia.org/wiki/Startup_Visa. Accessed 5 June 2012.

World Intellectual Property Organization: Technology transfer, intellectual property and effective university-industry partnerships: the Experience of China, India, Japan, Philippines, the Republic of Korea, Singapore and Thailand. http://www.wipo.int/freepublications/en/intproperty/928/wipo_pub_928.pdf. Accessed 27 May 2012, WIPO Publication no. 928E, Geneva, Switzerland; 2007.

WU Y. Indigenous innovation for sustainable growth in China: the next twenty years of reform and development. In: Garnaut R, Golley J, Song L, editors. Canberra: ANU E Press, The Australian National University; July 2010.

Zero2IPO: Global IPO marketplace (Q1 2012). http://www.pedaily.cn/Item.aspx?id=218729 (2012). Accessed 31 May 2012.

Zhang C, Zeng DZ, Mako W, Seward J. Promoting enterprise-led innovation in China. The World Bank; 2009.

Erratum to: Intellectual Capital: The World's Fastest Growing Asset Class

Erratum to:
Chapter 3 in: C. M. Gross, *Too Good To Fail*,
DOI 10.1007/978-3-319-00281-1_3

The correct version of the Fig 3.8 is updated below:

The online version of the original chapter can be found under DOI 10.1007/978-3-319-00281-1_3

C. M. Gross (✉)
Oxford Center for Innovation, Tekcapital Limited, New Road, Oxford OX1 1BY, UK
e-mail: cgross@tekcapital.com

C. M. Gross, *Too Good To Fail*, Management for Professionals,
DOI: 10.1007/978-3-319-00281-1_8,
© Springer International Publishing Switzerland 2013

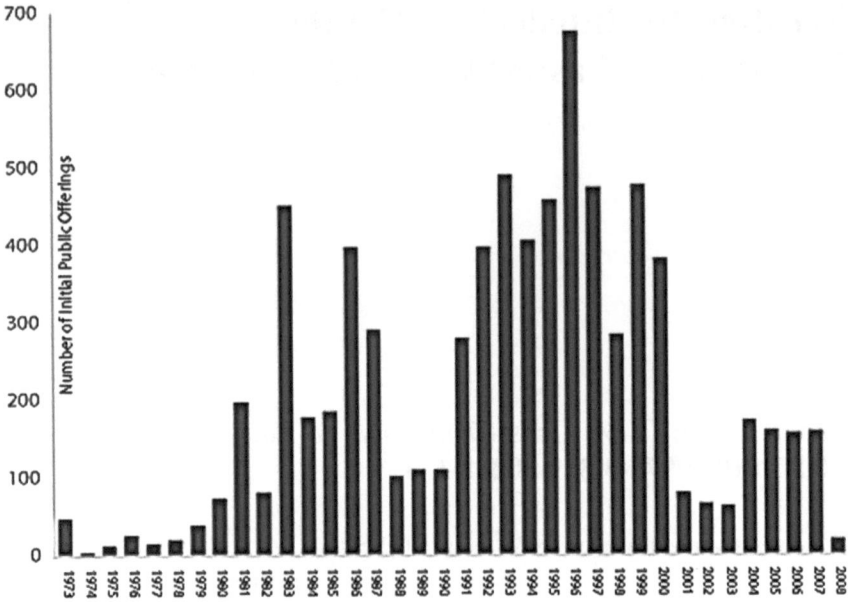

Fig. 3.8 US IPOs 1973–2008 (Ritholtz 2009, http://www.ritholtz.com/blog/2009/01/historical-chart-of-initial-public-offerings/)

About the Author

Clifford Gross is an executive with leadership experience in academia and commercial enterprises and is passionate about the development, transfer, and commercialization of intellectual property to create lasting value. He is an author/co-author of three prior books, The Right Fit, The New Idea Factory, and Technology Transfer for Entrepreneurs, and is a named inventor on 19 issued patents from his research.

Clifford founded two public companies where he served as CEO and Chairman. He also served as President and CEO of Innovacorp, an early-stage venture capital fund in Nova Scotia, Canada. Previously, he was acting director of the graduate program in biomechanics and ergonomics at New York University, Chairman of the Nelson Rockefeller Department of Biomechanics (NYCOM) at the New York Institute of Technology, and Research Professor at the University of South Florida.

He currently serves as the CEO and Chairman of Tekcapital Ltd. based in Oxford, U.K. and has completed more than 100 technology transfers from university and federal laboratories to a wide range of companies. He also serves on the board of directors of the Technology Transfer Society and the State University of New York at Empire State College. He received his Ph.D. from New York University and an M.B.A. from the University of Oxford.

C. M. Gross, *Too Good To Fail*, Management for Professionals,
DOI: 10.1007/978-3-319-00281-1,
© Springer International Publishing Switzerland 2013

Index

C. M. Gross, *Too Good To Fail*, Management for Professionals,
DOI: 10.1007/978-3-319-00281-1,
© Springer International Publishing Switzerland 2013

Research Perspectives in Couple Therapy